MICROBIOLOGY RESEARCH ADVANCES

THE HANDBOOK OF FOOD MICROBIOLOGICAL ANALYTICAL METHODS

MICROBIOLOGY RESEARCH ADVANCES

Additional books in this series can be found on Nova's website under the Series tab.

Additional e-books in this series can be found on Nova's website under the eBooks tab.

MICROBIOLOGY RESEARCH ADVANCES

THE HANDBOOK OF FOOD MICROBIOLOGICAL ANALYTICAL METHODS

ISMAIL MOHAMED AL BULUSHI

nova science publishers
New York

NOTICE TO THE READER

Library of Congress Cataloging-in-Publication Data

ISBN: 978-1-53612-473-6
Library of Congress Control Number: 2017957355

Published by Nova Science Publishers, Inc. † New York

CONTENTS

PREFACE

The Handbook of Food Microbiological Analytical Methods is a simple and step by step practical guide to conduct different food microbial analyses. It aims to provide and develop the food microbial practical skills in the users with simple steps to follow along with the theoretical explanation of the principle of each analytical method for a better understanding. This handbook includes 31 topics and experiments distributed in 5 chapters, namely basic microbial skills, enumeration of different microorganisms in foods, identification techniques and determination of microbial activities. Besides, the handbook includes useful sources in food microbial analyses such as the Food Microbiology E-sources and the primers for identification of common microbial pathogens.

Initially, this handbook demonstrates the basic safety skills for workers in food microbiology laboratories for a safe practical food microbiology session. The category of media and their preparations were discussed in detail to enable the users to prepare and provide a good nutritional source for microbial growth that can ensure the full recovery of microbes from foods. Enumeration techniques for specific bacteria such as marine bacteria, Gram-positive bacteria and lactic acid bacteria and other important groups are shown clearly through five simple steps. This handbook introduces for the first time a selective medium for the enumeration of Gram-positive bacteria. The thin agar layer technique as a technique for the enumeration and recovery of injured/stressed bacteria is elucidated with simple steps. Moreover, clear randomization techniques are provided to select the representative bacterial colonies for correct identification and further studies as well as clear guidelines in sampling for microbial analyses.

Common tests for the characterization of bacterial isolates such as Gram staining, potassium hydroxide, motility, spore-forming and sensitivity tests are also discussed with interpretation of the results. The principles and procedures of some routinely used identification techniques namely vitek, the sequencing of 16S rRNA gene and specific genus and specie primers for bacterial identification are provided. Enumeration and identification techniques of common pathogens and spoilers namely *Staphylococcus aureus*, Enterobacteriaceae, *Escherichia coli*, *Salmonella* spp., *Listeria monocytogenes*, *Vibrio parahaemolyticu*, *Bacillus cereus* and *Shewanella putrefaciens* enumeration and identification techniques are also covered.

Different tests for evaluation of the microbial potential to spoil foods, form biogenic amines in foods and produce bacteriocins are covered with simple and clear steps. These techniques were based on the International Standards such as ISO and U.S. Food and Drug Administration Standards and media's manufacture instructions. Moreover, this manual includes the specific specie primers of most common pathogens, so that it helps in molecular identification of these pathogens by using specific species primers techniques.

This handbook is needed to conduct microbial analyses to determine food microbial quality and safety, hygienic status and microbial potential to spoil food, loose food safety and produce antimicrobial activity. This handbook was prepared to be used by students and young researchers. Therefore, it can be used in universities as a practical manual in Biology, Microbiology, Food Microbiology,

Food Safety, Food Hygiene Courses, as well in food laboratories which determine food microbial safety and quality.

The uniqueness of this handbook is including topics that are rarely addressed in the current food microbiology manuals and handbooks such as the enumeration of special bacterial groups such as marine and Gram-positive bacteria, introducing; a selective medium for enumeration of Gram-positive bacteria from marine sources, a technique for bacterial colonies randomization, a technique for recovery of injured/stressed bacteria, techniques to study bacterial potential such as spoiling foods, forming biogenic amines and producing antimicrobial activity and providing the specific specie primers pairs for common pathogens. It is expected that this handbook will be a good practical guide for students and researchers as well.

ACKNOWLEDGMENTS

I greatly thank Dr. Mohamed Essa for his kind advice and useful guidelines to publish this handbook. Dr. Essa thankfully provided the main ideas required to submit the manuscript of this handbook, in addition to his kind recommendations for potential publishers. Indeed, his kind help was one of the means which made it possible for this handbook to be published.

I also greatly thank Prof. Mohamed Shafiur Rahman for his kind advice for the requirements and procedures of book publication.

About the Author

Ismail Mohamed Al Bulushi, PhD

Department of Food Science and Nutrition
College of Agricultural and Marine Sciences
Sultan Qaboos University, Oman

Ismail Mohamed Al Bulushi received his PhD in Food Microbiology from the University of Queensland (Australia) in 2009. Dr. Al Bulushi jointed Sultan Qaboos University (SQU), Oman in 1989 as a Technician. Currently he is working as an Assistant Professor in the Department of Food Science and Nutrition at SQU. He is teaching Food Microbiology, Public Sanitation, Fish and Fish Products, Meat and Meat Products courses. Dr. Al Bulushi developed many practical experiments for his teaching courses and validated Gram-positive bacterial enumeration medium and incubation conditions. He supervised many undergraduate and graduate thesis projects in several areas of Food Sciences such as the chemical and microbial characteristics of date fruits, the microbial floral composition of traditional fish products, halotolerant bacterial count in traditional salted fish products, development of many fish products and the antimicrobial activity of some plants. He organized several conferences, symposia, workshops, lectures and extension activities in community services. Dr. Al Bulushi provided many national and international consultancies in several Food Science areas. His research areas focus on the microbial, chemical and physical characteristics of traditional foods, marine fish toxicity, freshness, safety, spoilage and processing, marine fish microbial flora's composition with special references to the Gram-positive bacterial population and microbial spoilage, biogenic amine formation and antimicrobial activities potential. Dr. Ismail published some papers and articles in international journals, magazines and newspapers and participated in many international conferences and workshops in Oman and overseas.

Address correspondence to:
isab@squ.edu.om and ismailalbulushi2014@gmail.com.

Chapter 1

BASIC MICROBIAL SKILLS

INTRODUCTION

This chapter aims to build-up the basic microbial skills in the student, researchers and any microbial laboratory user. It exposes the users to the different essential skills that are needed in any microbial laboratory. In the beginning of this chapter, the users are introduced to the different risks such as fires, microbial and chemical hazards that might risk the user's life. The users are guided to the proper controlling means that reduce the risks of these hazards. Writing a proper scientific report is one of the challenges facing students and researchers, thus in this chapter, students and researchers are provided with the scientific report format. They are guided as to what should be written in each part of this report such as how to present the results, how the results should be discussed, what should be included in the conclusion and how to insert the references in the text and in the reference list. This chapter makes the user of a microbial laboratory familiar with the different types of microbial media such as selective and non-selective media. It trains the user on how to prepare the microbial media with very simple steps to follow. This chapter also provides the base of food sample dilution which enables the user to dilute the sample in order to make the microbial colonies within the counting range. In order to select the representative colonies from colonies population, this chapter introduces the Harrison's disc, a simple technique that can ensure the accurate and representative selection for any further use of these colonies. Recovery of injured or stressed microbes is one factor which reduces the microbial counts from selective media in frozen and dried foods. Thus, this chapter provides a thin agar layer, a technique which improves the recovery of injured microbes enumerated on selective media by encouraging their growth, first on the agar layer in selective media. Microbial initial characterization, which can be used for their identification are determined with simple tests that can be followed by any user even without a biological background. The principle of these tests is provided to help the user to understand and interpret the results of this test. The chapter provides proper guidelines in sampling for microbial analyses which can be applied in routine analyses and research works. These guidelines are based on international standards. Finally, the chapter provides the proper means for microbial short and long storage as well as their recovery and maintenance.

WORK IN FOOD MICROBIOLOGY LABORATORY

Aims

- To know the safety precautions in the food microbial laboratory.
- To train on the cleaning and sterilization of glassware, culture media, etc.
- To familiarize with the equipment and apparatus used in microbial work.

Introduction

Working in a microbiological laboratory exposes the workers to many hazards such as chemicals, flammables and explosive substances, microorganisms, ultraviolet and other radiations. It should always be assumed that these hazards are capable of harming and causing diseases. Understanding the safety procedures could protect the workers from different hazards and provide a safe environment for microbial work. A microbial laboratory is equipped with different equipment such as balances/scales, autoclave, biosafety cabinet, laminar flow cabinet, stomacher, etc. Using this equipment properly will ensure the quality of the microbial works and maintain this equipment for long use. All microbial works are run under clean, sterilized and aseptic environments. Training on these procedures will provide good results and minimize the risk of microbial hazards in the laboratory.

Personal Protection

- A laboratory coat must be worn in all laboratory works. Caution: no student is allowed to work in the laboratory without a lab coat.
- A used lab coat must be autoclaved before being sent for laundering. Do not wash your microbiological laboratory lab coat at home.
- A microbial laboratory lab coat must not be borrowed or lent. This will prevent cross- contamination.
- Chatting and unnecessary talk should be avoided during work.
- No food or drink is allowed to be consumed or stored in the microbial laboratory except food samples.
- The working area must be sterilized before and after work.
- All items must be labeled.

- Great care should be taken when using ultraviolet light. Exposure to ultraviolet light might damage your body.
- Microbial cultures, plates, and used apparatus should not be exposed to air.
- Bunsen flames must be controlled and never left.
- In all personal accidents, medical and lab supervisor help should be sought.
- On completion, hands must be washed with antimicrobial detergents.
- All work in pathogens and pure cultures are carried out in biosafety cabinets.

Accidents and Fires

- In case of fire, **protect yourself first**, run to an assembly point.
- Never leave a Bunsen flame close to alcohol, cotton, paper towel, personal belongings or unattended.
- Learn how to contain a small fire.
- Learn how to use fire extinguishers.
- Close the doors and leave the building immediately in case of life-threaten fires.
- Find the first aid kit in the laboratory.
- A spilled culture should be flooded with suitable disinfectant solutions for 15- 30 min. before cleaning up.
- Broken glass, sharps, contaminated glass or metal items should be placed in an autoclavable sharps container.

Laboratory Practices

- Never throw contaminated petri dishes in the general garbage bin, the laboratory provides biohazard bins.

- No microbial item should be left in the sink or working area after laboratory completion. Wash your stuff after completion.
- Inoculating needles and loops must be sterilized by heating in a Bunsen flame until red hot along the entire length of the needle or loop.

Familiarization with Common Laboratory Equipment

- Autoclave
- Stomacher
- Dispenser
- Biosafety cabinet
- Fire extinguisher
- First aid kit
- Incubator
- Refrigerator
- Optical microscope
- Water bath
- Vertex
- Oven

Cleaning and Sterilization of Glassware and Apparatus

- In microbial works, usually disposable/single use items are used.
- All used microbial items must be autoclaved before cleaning.
- Used pipettes are placed in a pipette jar containing disinfectant solution.
- Plastic Petri dishes, cultures, bottles containing microorganisms are sterilized in an autoclave at 121°C for 30 min.

- Glassware is autoclaved in an air oven at 160°C for two hours (used only for heat- stand materials) (Al Bulushi 2016).

FOOD MICROBIOLOGY PRACTICAL REPORT FORMAT

Aim

To know how to write a laboratory report.

Introduction

Writing a scientific report following the recommended format will help the student/researcher to present his/her results and show the importance, value and quality of the experiment results in a simple and understandable way. Despite the nature of the experiment, all reports should meet the standard format which includes introduction, aim/objective, materials and methods, results, discussion, conclusion and references. The following guidelines are based on a laboratory manual (Al Bulushi 2016).

Format of Laboratory Report

- Experiment title:
- Student Name:
- Student ID:
- Date:
- A short background about the topic, definitions, importance/value of the topic, factors affecting, etc. cited either from textbooks, reports, scientific papers, etc.
- What is the problem: why are you doing this experiment?

- Aims: how will you solve this problem?

Materials and Methods

- What is your sample and where and how did you get it?
- Which items/apparatus did you use?
- Which methods did you use?

Results

- What did you get from this experiment?
- Present your results in tables/ graphs/ pies. Write a few lines about the results.
- Compare between samples/ groups e.g., sample A is lower than sample B.

Discussion

- Explain your results. State the reasons. For instance: why are the values high or low, why sample A is different from B: what do you think are the reasons?
- Compare your results with some published results to provide a scientific ground for your work.

Conclusion

- Summarize the main experiment results, what did you find?

- What do these values mean: what can you learn/benefit from these results?
- Usually, no numbers or values are written in the conclusion.
- Conclusion is expressed in a few simple, clear words. Do not write background or new information in conclusion.

References

References can be:

- Papers in a journal
- Chapters in a textbook
- Text book
- Conference proceeding
- Website
- Report
- Thesis
- Personal communication
- etc.

There is no standard style to write the references. Each journal has its own style.

Examples below are the style followed by the International Journal of Food Microbiology

Paper in a Journal

Al Bulushi, I., Poole, S., Deeth, H., Dykes, G., 2008. Quantitative assessment of the total and Gram-positive aerobic bacteria in fresh and ambient-temperature-stored sub-tropical marine fish. World Journal of Microbiology and Biotechnology 24, 1867-1875

Chapter in a Textbook

Bonde, G., 1981. *Bacillus* from marine habitats: allocation to phena established by numerical techniques. In: R. Berkeley, M. Goodfellow (Eds.), The aerobic endospore-forming bacteria: classification and identification. Academic Press, Kent., 181-215.

Textbook

Whitman K. 2004. Finfish and Shellfish Bacteriology. Iowa State Press, Iowa.

MEDIA AND DILUTION SOLUTIONS

Aims

1. To familiarize with microbial media and dilution solutions.
2. To learn how to prepare microbial media and dilution solutions.
3. To learn how to use an autoclave and other materials in media preparation.

Introduction

A microbial medium is any nutrient liquid or solid that can be used in the laboratory for the growth of microorganisms. A microbial medium is composed of the necessary requirements for growth which vary according to the microorganisms but include water, nitrogen-containing compounds (e.g., peptides, proteins, amino acids), energy source (e.g., carbohydrates, peptides, amino acids, proteins) and accessory growth factors such as time, temperature, pH, water and oxygen.

As microorganisms' nutritional requirements vary, there is no single medium capable of supporting the growth of all microorganisms. Therefore, a specific microbial nutrition requirement can be added to a *basal medium.* Media are available in dehydrated forms where water is added to the powder and sterilized according to manufacturer's instructions or can be prepared in the laboratory from different components. The pH of the media necessary needs to be adjusted before autoclaving as the growth and metabolism of microorganisms are affected by pH.

Based on selectivity, media can be divided into non-selective media and selective media. Non-selective media are those which are used for the cultivation of most bacteria such as nutrient agar, standard plate count agar and tryptone soya agar. Selective media are those containing specific substances which support the cultivation of some bacteria and inhibit the growth of others. Selective media have selectivity either to a specific group of bacteria such as De man, Rogosa, Sharpe (MRS agar) which is used to isolate lactic acid bacteria or to a single species such as Baird-Parker agar which is used to isolate *Staphylococcus aureus* (Schoeller and Ingham 2001).

A microbial sample is diluted at different strengths to get a microbial count within the range of 30-300 colonies. The amount of samples and the number of dilutions needing to be prepared depends on the methods which are followed, nature of microorganisms being targeted and the initial quality of food. For pathogens, the standards recommend to withdraw 25g of sample and the presence and absence of pathogens is reported in 25g. For high quality food, low dilutions are usually prepared.

Many diluents (substances added to prepare dilutions) such as distilled water, most recovery diluent (MRD) and Ringer's solution are used, nevertheless, ISO standards in recent years have started to specify 0.1% peptone in 0.85% sodium chloride as the normal general purpose diluent (ISO/DIS 6649). A suitable diluent for an osmophilic (microorganisms that can grow at a relatively higher osmotic environment than that is needed for other microorganisms) count is sterile 20% sucrose solution and for a halophilic (microorganisms that can grow at a relatively higher salt content than is needed for other microorganisms) count is 15% sodium chloride.

Assignment

- Preparation of 250 ml standard plate count agar.
- Preparation of 45 ml and 9 ml of 0.1% peptone in 0.85% sodium chloride diluent.

Materials

- Dehydrated standard plate count agar
- Peptone
- Bacteriological grade sodium chloride
- Media bottle, 100 ml, 250ml
- Dilution bottle, 45 ml
- Test tubes, 10 ml
- Top loading balance
- Autoclave
- Sterilization indicator strep
- Refrigerator at 4°C
- White tape
- Marker
- Petri dish
- Beakers 100 ml, 250 ml
- Spatula, spoon
- Distilled water, 500 ml
- Measuring cylinder, 100 ml, 250 ml

- Disinfectant
- Tissues/papers towel
- Pipette, 10 ml

Methods

Standard Plate Count Agar

1. Prepare 250 ml standard plate count agar according to the instructions of the manufacturer. Mix well to dissolve. No need for preheating if the medium is not dispensed.
   Loosen the covers to release the air inside the containers built up during sterilization and sterilize the medium at 121°C for 15 min.
2. Bring the temperature of the medium to about 50°C, sterilize your working area with disinfectant and pour about 12 ml into petri dish. This should make 18 plates.
   Slowly move the medium to cover the bottom of plate, do not shake. Avoid contaminations during the pouring of plates.
3. After solidification, label your plates and store the plates at 4°C.

Diluent

1. Prepare 120 ml of 0.85% sodium chloride.
2. Prepare 100 ml 0.1% peptone in 0.85% sodium chloride. Mix well.
3. Dispense 45 ml of prepared solution into a medium bottle and label it.
4. Dispense 9 ml of prepared solution into a test tube and label it.
5. Repeat steps 2 and 4 above.

Maximum recovery diluent (MRD) is available commercially.

Randomization and Isolation Techniques of Bacterial Colonies

Aims

1. To familiarize with the random selection technique of bacterial colony.
2. To know the bacterial colony's isolation and purification techniques.

Introduction

Bacteria are isolated and purified in order to be identified and their metabolic activities to be studied. This involves selecting and purifying a single representative colony in a non-selective medium and increasing the agar content to 2% to prevent the colony from spreading. Selecting a representative bacterial colony among hundreds of colonies for identification and studies always raises concerns about the justification for such selection modes due to different colonies' characteristics. Representative samples of colonies cannot be secured by uncontrolled picking from a Petri dish and various methods have been developed to ensure that a truly random sample of colonies is selected. Among the methods is the selection of every colony occurring either in a single sector or in opposite sectors of the plate. This can be achieved by using Harrison's disc (Harrison 1938, cited by Harrigan 1998). The petri dish to be examined is placed concentrically on the disc. First, all colonies occurring in the areas marked 1 are selected. If areas marked 1 do not provide sufficient numbers, all colonies occurring in the areas marked 2 are selected. Areas marked 3 and 4 can be used if a sufficient number of colonies has not been obtained in areas 1 and 2. All colonies lying over the lines should be ignored.

Materials

- Harrison's disc template (can be prepared manually according to specifications provided by Harrigan, 1998)
- Petri dish of about 100 colonies
- Pre-prepared plates of 2% agar tryptone soya agar. The base medium is prepared according to the instructions of the manufacturer.
- Incubator at 25°C
- Loops

Methods

- Place a plate with 10-100 colonies on the Harrison's disc template.
- In the areas marked 1, select 3 single colonies (normally 10 to 25 colonies are selected). Use other areas to get sufficient colonies.
- Streak each single colony in a 2% agar standard plate count agar as following:
  - Spread a small inoculum over one-half of the plate toward the edge.
  - Sterilize the loop and make one vertical stroke from the end of last stroke in step 1.
  - Sterilize the loop and make some horizontal strokes from the end of last stroke in step 2.
  - Incubate the plates at 25°C for 3 days.
  - To purify the bacterial colony further, streak single the colony in a 2% agar standard plate count agar as described in step 3.

- Repeat step 4 until pure colonies are obtained which can be verified by the colour and size of the colonies.
- Keep the pure colonies at 4°C for bacterial identification.

Thin Agar Layer Technique

Aims

- To recover the injured/stressed microorganisms in foods.
- To familiarize with thin agar layer technique.

Introduction

Upon thermal processing or the addition of chemical additives or lowering the pH of food, many vegetative microorganisms including pathogens suffer from injury or stress which makes their enumeration in food difficult and less efficient by following the normal techniques followed in fresh foods. In this respect, if injured or stressed bacteria are grown in selective media, the chemical inhibitors in these media increase bacterial cells injuries and results in very low recovery. In fact, many techniques have been developed to maximize injured bacterial recovery.

Principle

The thin agar layer (TAL) technique was developed by Kang and Fung (2000) and has been widely used and found to be efficient in the recovery of injured pathogenic bacteria on selective media. The thin agar layer technique involves overlaying the selective media with a thin layer, 14 ml of non-selective media. Initially, all injured bacteria can grow on the thin layer of non-selective media, and then they penetrate the selective media. Based on the nature of chemical inhibitors and target bacteria, the growth of some bacteria is inhibited by these chemicals, whereas the target bacteria, resistant to these chemicals will grow and can be easily enumerated.

The enumeration of *Staphylococcus aureus* on Baird Parker agar will be used as an example of TAL application in the recovery of injured/stressed microbes.

Materials

- Dried fish
- Pre-prepared plates of Baird Parker agar enriched with egg yolk and potassium tellurite and overlaid with two layers of 7 ml standard plate count agar, SPCA (can be done by pouring 7 ml of SPCA on the surface of dried plates of BPA. Then after solidification, pouring another 7 ml of SPCA). The base medium is prepared according to the instructions of the manufacturer.
- Pre-prepared plates of 2% agar standard plate count agar
- 225 ml, 9 ml of MRD
- Stomacher
- Stomacher bag
- Scissors, knife, forceps
- Petri dish
- Top loading balance
- Stop watch
- Tibs 1 ml
- Tibs dispenser
- Glass spreader
- Incubator at 37°C

- 70% ethanol
- Bunsen flame
- Beaker 250 ml
- Tissue
- Vertex
- Colony counter
- Harrison's disc template
- Loops
- Chopping board
- Staphytect plus system

Methods

Enumeration of Staphylococcus aureus

Sampling

- Aseptically remove about 50g of dried fish and cut into small pieces.
- Put a stomacher bag in the beaker and weigh 25g of the sample.
- Transfer 225 ml of MRD into the stomacher bag and blend for 1 min.

Diluting

Step 3 prepares dilution of 10^{-1}. Prepare further dilutions (based on the handling of food e.g., good handled food needs low dilutions, mishandled foods needs high dilutions) by transferring 1 ml of the previous dilution into 9 ml MRD. For example, to prepare a dilution of 10^{-2}, transfer 1 ml of 10^{-1} to 9 ml MRD.

Plating

- Select 3 appropriate dilutions to be plated.
- Transfer 0.1 ml from each dilution into 2 plates of Baird Parker agar prepared with TAL.
- Using the glass spreader, spread the samples over the plates.
- Label 2 plates of Baird Parker agar as control.

Incubating

Incubate the Baird Parker agar plates inverted for 48 ± 2 h at 37°C.

Counting/Enumerating

Count only the black convex and shiny colonies with narrow white margins that are surrounded by clear zones. Insert the colonies in the following table:

Sample	Dilutions									
	10^{-1}		10^{-2}		10^{-3}		10^{-4}		10^{-5}	
	Plates		Plates		Plate		Plates		Plates	
	1	2	1	2	1	2	1	2	1	2

Reporting the Bacterial Count

- Consider the plates with 30- 300 colonies for reporting of S. *aureus* count in the sample.
- Report *S. aureus* count as log cfu/g.

Confirmation of Staphylococcus aureus

- Using the Harrisons disc template select 3-5 colonies randomly from a plate containing 100 colonies.
- Streak each colony on a 2% agar standard plate count agar.

- Incubate the plates at for 48 ± 2 h at 37°C.
- Confirm *S. aureus* using Staphytect plus system which detects coagulase enzyme according to the manufacturer's instructions.

Result

Staphylococcus aureus is coagulase positive.

Characterization of Bacterial Isolates

Aim

To familiarize primarily with test principles and procedures which are used to characterize bacterial isolates from foods.

Introduction

Bacterial isolate can be characterized using colony and cell morphology, Gram stain, Gram reaction, motility test, catalase test, oxidase test, spore-forming test, oxidative-fermentative test, O/129 sensitivity, reduction of trimethylamine oxide, production of hydrogen sulfide, hydrolysis of Tween 20 and decarboxylation of arginine. These characteristics could be used to presumptively identify bacteria to the genera level. Fortunately, some of these tests are included in many commercial bacterial identification systems. Following are the principles and procedures of these tests.

Morphology of Colony

- Shape: circular, irregular, rhizoid
- Pigment: white, yellow, orange, pink

Morphology of Cell

- Shape: spherical (cocci), rod shaped (bacilli), curved (comma)
- Appearance: cluster, chain (two or more cells), tetrads

Gram Staining

Principle

Gram staining which was devised in 1844 by a Danish bacteriologist, Hans Christian Gram is based on the ability of a bacterial cell to retain crystal violet dye when the bacterial smear is decolorized. The bacteria are initially stained with crystal violet. Iodine is added to fix the crystal violet. The stain is then decolourized with a mixture of acetone and ethanol. The acetone- ethanol mixture dissolves the lipid layer from Gram-negative cells walls allowing leaching of the crystal violet from the cells. When safranin is added, the decolourized Gram-negative cell stains pink. In contrast, the acetone-ethanol mixture dehydrates Gram-positive cell walls resulting in the purple crystal violet remaining.

Factors affecting Gram staining results:

- Overheating during drying of the bacterial smear will cause all cells to appear Gram- negative bacteria.
- Old cultures can cause cells to appear Gram-negative bacteria.
- If cells are taken from liquid or selective media, cells will appear Gram-negative bacteria
- Weakly fixing by iodine, either due to use of an old reagent or improper flooding, the cells will appear
- Gram-negative bacteria. The conventional preparation of iodine is relatively unstable and may lose up to 60% of the available iodine in 30 days at 25°C.

- Decolorizing with improper agents will cause all cells to appear Gram-positive bacteria.

Determination of Gram Reaction

Principle

Gram reaction, whether bacteria are negative or positive to Gram, is determined by 3% potassium hydroxide (Halebian et al. 198). The 3% KOH test is based on differences in the chemistry of the bacterial cell walls. The cell wall of Gram-positive bacteria composes many layers of peptidoglycan and that of Gram-negative bacteria contain few layers of peptidoglycan. The cell wall of Gram-negative bacteria is easily disrupted when exposed to diluted alkali solutions. When the cell walls are disrupted, the suspension in KOH becomes viscous due to the release of relatively unfragmented threads of DNA. Potassium hydroxide has no detectable effect on the cell walls of Gram-positive bacteria.

Catalase Test

Principle

Most bacteria possess the enzyme catalase. This enzyme breaks down hydrogen peroxide (H_2O_2) to water and oxygen molecules. The reaction can be detected by bubbling. The lactic acid bacteria (including *Streptococcus* spp., *Lactococcus* spp., *Leuconostoc* spp., *Lactobacillus* spp.) however, do not normally produce a detectable catalase. This test discriminates those which produce this enzyme from those which do not. It is used to easily differentiate lactic acid bacteria from other cocci such as *Micrococcus* spp. and *Staphylococcus* spp.

Oxidase Test

Principle

This test detects a microorganism's abilities to produce cytochrome oxidase. Tetramethyl p- phenylene diamine is oxidized in the presence of oxygen to form dark purple indophenol. The test can differentiate *Micrococcus* spp. which possess oxidase from *Staphylococcus* spp. which do not produce oxidase except in one species, *Staph. sciuri.*

Motility Test

Principle

Bacteria move with the use of flagella (20 nm in diameter, 15-20μm long) which come from the surface of the cell, rotation of the cell by axial filament and the gliding movement of the cells over the surface without the aid of flagella. In action, the filament rotates from 200 to more than 1000 revolutions per second. Bacterial motility can be detected by stabbing the bacteria in sulfide indole motile medium and examining the growth mode, as this medium has been recommended by USA FDA to detect the motility in bacteria (Cited by Himedia 2015). Motile bacteria grow away from the stabbing line. Another method is using the hanging drop technique in which bacterial movement can be seen under a compound microscope.

Spore-Forming Test

Principle

Under harsh conditions such as high temperatures, lack of nutrients, low water activity and chemical and antimicrobial treatments, some genera such as *Bacillus, Alicyclobacteria, clostridium* and *Desulfotomaculum* which implicate in fish spoilage and foodborne diseases produce endospores (inside a cell and one spore per cell).

Endo- spores highly resist the harsh and stressful conditions. The spores are metabolically inactive or dormant and can remain dormant for years, but can emerge as vegetative cells in a suitable environment. Sporulation is followed by a germination step which is considered as a metabolically degradative process. Spores are stained with malachite green under a steam bath which softens the sturdy and durable outer layer of the cell and allows malachite green to penetrate and bind within the spore. After washing, the stain comes out of the cell walls of the vegetative cells making them colourless but not from the spore wall. When the bacterial smear is counterstained with safranin, the vegetative cells appear red/pink whereas, the spore remain green.

Oxidative/Fermentative Test (OF Test)

Principle

Bacterial metabolisms of carbohydrates can be described by two processes; fermentation and oxidation. In fermentation, the glucose molecule first is phosphorylated and then split into triose molecules which undergo further changes. This metabolism is independent of the oxygen process. The oxidation involves the splitting of glucose molecules to aldehyde group. This process is strictly aerobic. Glucose metabolisms can be detected by changes in the colour of the medium which contains a pH indicator, bromthymol blue. Bromthymol blue produces a green colour at pH 7.1. When glucose is fermented, the pH of the medium becomes acidic causing the colour of bromthymol blue to change from green to yellow (Hugh and Leifson 1953).

O/129 Sensitivity

Principle

The pteridine derivative (2, 4-diamino-6, 7-di-isopropyl pteridine phosphate) specifically inhibits the growth of *Vibrio* spp. and other bacteria. Resistance can be detected by placing discs of the reagent on plates previously inoculated with the target microorganisms. A zone of inhibition indicates sensitivity to O/129 (Ibarra and Alvarado 2007).

Morphology of Colonies

Materials

- Fresh colonies (24-48 hour-colonies of your purified colonies)
- Colony counter

Method

- Select a plate with 10-100 colonies and check the colonies under a colony counter.
- Describe the shape of colonies as circular, irregular, rhizoid.
- Describe the colour of colonies as white, yellow, orange, pink.

Gram Staining and Morphology of Cells

Materials

- Fresh colonies (24-48 hour-colonies of your purified colonies)
- Slide, 2
- Bunsen flame
- Loop
- Distilled water
- Gram stain set (crystal violet, fresh iodine solution, decolourizer: (acetone: ethanol, 1:1, safranin)
- Tray for stain waste
- Hair dryer
- Compound microscope
- Tissues

- Immersion oil

Preparation of Bacterial Smear

Method

- Place two clean slides on the tissue paper.
- With a loop, place a small drop of water on one side of the slide.
- Sterilize the loop until redness and allow to cool.
- Pick off a little of the bacterial growth. Avoid transferring any agar medium.
- Emulsify the growth and smear it evenly over the area of the slide. Aim to obtain a thin and spread smear.
- Dry the bacterial with a hair dryer. Avoid heating the smear.

Gram Staining

Method

- Flood the slide with crystal violet for one min.
- Wash the slide with a gentle and indirect stream of water, dry. Do not discard the stain in the sink, use a stain waste container.
- Flood with iodine for one min.
- Repeat the washing and drying step.
- Decolourize the slide with acetone-ethanol solution.
- Repeat the washing and drying step.
- Flood the slide with safranin for one min.
- Repeat the washing and drying step.
- Place the slide under microscope and check the appearance of cells first at × 10 objective.

- Put a drop of immersion oil on the slide and check at × 100 objective.
- Determine the Gram reaction and cells morphology.

Result

- Gram-positive bacteria will appear blue or violet
- Gram- negative bacteria will appear pinkish red.
- Describe the shape of cells as spherical (cocci), rod shaped (bacilli), curved (comma).
- Describe the appearance of the cells as cluster, chain (two or more cells), tetrads.
- Repeat the test three times to confirm the results.

Determination of Gram Reaction by 3% KOH

Materials

- Fresh colonies (24-48 hour-colonies of your purified colonies)
- Freshly prepared 3% KOH solution
- Slide
- Loop
- Tissues
- Medicine dropper

Method

- Place a clean slide on tissue paper.
- Put one drop of 3% KOH solution on the slide
- Pick off a little of the bacterial growth.
- Stir the bacterial growth in a circular motion in the KOH solution for 30 sec.

- Raise the loop slowly and observe texture of bacterial growth in KOH solution.

Result

- Viscous texture with stringing occurred within 30 sec.: Gram-negative bacteria.
- Watery texture with absence of stringing: Gram-positive bacteria.
- Note: the result should be considered within 30 sec of stirring.
- Repeat the test three times to confirm the results.

Catalase Test

Materials

- Fresh colonies (24-48 hour-colonies of your purified colonies)
- Freshly prepared 3% H_2O_2 solution
- Slide
- Loop
- Tissues
- Medicine dropper

Method

- Place a clean slide on tissue paper.
- Put one drop of 3% H_2O_2 solution on the slide
- Pick off a little of the bacterial growth.
- Stir the bacterial growth in a circular motion in the H_2O_2 solution for one min.

Result

- Liberation of free oxygen as gas bubbles: catalase positive.
- Absence of gas bubbles: catalase negative.
- Repeat the test three times to confirm the results.

Oxidase Test

Materials

- Fresh colonies (24-48 hour-colonies of your purified colonies)
- Oxidase sticks

Method

- Touch the oxidase stick to the surface of colony thoroughly.
- Observe the colour changes in oxidase stick.

Result

- Purple colouration within 5-10 sec: oxidase positive
- Absence of purple colouration: oxidase negative
- Repeat the test three times to confirm the results.

Motility Test by Sulfide Indole Motile Medium

Materials

- Fresh colonies (24-48 hour-colonies of your purified colonies)
- 5 ml of SIM medium
- Sterile needle

- Incubator at 37°C

Method

- Using a sterile needle, pick a colony.
- Stab SIM medium to within 1 cm of the bottom of the tube. Keep the needle in the same line as it is inserted while removing it from the medium.
- Incubate the medium at 37°C for 24 hr.
- Observe the growth mode.

Result:

- Motile: Diffusion of growth away from the stabbing line.
- Non-motile: Growth on the stabbing.
- Repeat the test three times to confirm the results.

Spore-Forming Test

Materials

- Fresh colonies (24-48 hour-colonies of your purified colonies) - 5% solution of malachite green
- Safranin
- Fresh Gram iodine
- Slide
- Steam bath (metal rack on beaker with boiling water)
- Compound microscope
- Tissue
- Medicine dropper
- Hair dryer

- Tray for stain waste
- Fume hood
- Stop watch
- Loop
- Bunsen flame
- Immersion oil

Method

- Prepare the bacterial smear as it is described in Gram staining.
- In the fume hood, prepare a steam bath.
- Place the bacterial smear on the rack over the steam bath.
- While steaming, flood the bacterial smear with malachite solution for 5 min.
- Avoid the vapor of stain.
- Avoid dryness of the smear by adding the stain while steaming.
- Cool the slide.
- Wash really well then dry.
- Flood the slide with fresh Gram iodine solution for one min.
- Wash and dry.
- Flood the slide with safranin for one min.
- Wash and dry.
- Put a drop of immersion oil on the slide and examine at 100× objective.
- Examine the colour of the cells

Result

- Red/pink: vegetative cells
- Green: spores
- Repeat the test three times to confirm the results.

Oxidative/Fermentative Test

Materials

- Hugh and Leifson (1953) medium:
  - 0.2% Peptone
  - 0.5% NaCl
  - 0.03% K_2HPO_4
  - 0.2% agar
  - 0.003% bromthymol blue
  - 1% glucose (or any sugar in question)
  - pH 7.1
- Fresh colonies (24-48 hour-colonies of your purified colonies)
- Sterile mineral oil
- Loop
- Bunsen flame
- Incubator at 25°C

Methods

- Preparation of Hugh and Leifson (1953) medium:
- Mix all ingredients except glucose in water and adjust the pH.
- Heat the medium until dissolving all ingredients.
- Dispense 5ml of the medium into 10 ml test tube.
- Autoclave at 121°C for 15 min. Cool the medium to about 45°C.
- Prepare 10% glucose solution. Sterilize glucose solution through membrane filtration (pore size of 0.45μm).
- Add 0.5ml of sterile glucose solution to 5ml of Hugh and Leifson (1953) medium. Mix by inverting, do not shake.
- Oxidative/ fermentative test

- Inoculate two test tubes of Hugh and Leifson (1953) medium with the bacterial growth.
- Seal one test tube with 0.3ml of sterile mineral oil.
- Incubate the medium at 25°C for 4 days.
- Observe the changes in the colour.

Result

- Both tubes are green (no change): neutral, cannot use carbohydrates.
- Both tubes are yellow: fermentative.
- Open tube is yellow, sealed tube is green: oxidative
- Repeat the test three times to confirm the results.

O/129 Sensitivity

Materials

- Bacteria grown in broth (24-48 hours culture)
- Prepoured plates of standard plate count agar
- O/129 discs (10 μg and 150 μg) (Roberts and Greenwood 2003)
- Glass spreader
- 9 ml diluent (0.1% peptone in 0.85% NaCl)
- Forceps
- Incubator at 25°C
- Tibs 1ml

Method

- Dilute the bacterial growth to 10^{-6}
- Inoculate plates of standard count agar with 0.1 ml of 10^{-6} bacterial suspension, spread over the plate.

- Using the sterile forceps, put three discs of O/129 on the plate.
- Incubate the plates at 25°C for three days.
- Examine the plate for inhibition of bacterial growth around the discs.

Result

- Clear zones around the disc: sensitive
- No zones around the disc: resistance
- Repeat the test three times to confirm the results

Microbial Culture Storage and Maintenance

Aim

To familiarize with storage and maintenance techniques of pure microbial culture.

Introduction

Pure microbial culture can be purchased from national and international microbial collection centers. Researchers and students can also collect his/her own culture from routine experimental works and use it in future studies. The main requirement for culture collection is the purity of the culture which indicates that it is a single microbe. The microbial culture can be stored temporarily for a short time e.g., weeks on a suitable nutrient in slant and plates, and for a long time e.g., years on beads. Microbes can be temporarily stored at chilling temperatures and for long storage at freezing temperature such as −80°C (Al Bulushi 2009). As the microbe culture might lose its activity due to growth cycle sequence nutrition limitation in the medium and its identity due to mutation if it is kept for a long time, regular sub-culturing of the microbe culture is essential to keep it active and its identity.

Materials

- Young microbial culture obtained from a microbial experiment and purified as previously indicated
- Slant of tryptone soya agar
- Tryptone soya broth
- Tryptone soya agar
- Loop
- Vial of beads for microbial storage
- Incubator

Methods

Microbial Culture Storage

Temporary Storage

- Strike a small touch of microbial growth from a pure culture on the surface of tryptone soya gar slant.
- Incubate the slant at the optimum growth temperature for struck culture.
- After incubation, keep the culture at chilling temperature.
- Discard the culture immediately if the morphology changes.
- Long period storage
- Transfer a heavy loop-full of microbial growth into a vial of beads for microbial storage.
- Mix well the inoculant with the beads.

- Allow the microbe to settle on the beads.
- Discard the liquid surrounding the beads.
- Keep the vial at−80°C.

Microbial Culture Maintenance

- Regularly, the microbial culture is needed to be activated in order to maintain its activity and identity. This can be done as follows:
  - Transfer a bead or a touch of microbe to tryptone soya broth.
  - Incubate at the optimum growth temperature for the microbe.
  - Check the medium turbidity which indicates the degree of microbial growth.
  - If the microbe does not change the medium, a longer incubation or a heavier inoculant is needed.
  - Transfer a loop-touch of microbe growth from broth and strike it on the surface of tryptone soya agar.
  - Incubate at the optimum growth temperature for the microbe.

Store the microbe for temporary or long storage as described in step 1.

GUIDELINES IN SAMPLING FOR MICROBIAL ANALYSES

Aims

- To know the sample size and unit for different microbial enumeration and detection.

- To familiarize with sampling process for microbial analyses.
- To understand sample handling conditions.

Introduction

The sample is a small unit of a food or unit that can represent the whole food in its chemical, physical and microbial characteristics. Food is sampled to evaluate its safety, quality and compliance with food regulations. Therefore, a wise and proper withdrawing of this sample from the whole food should be conducted to accurately evaluate the food safety and quality and other concerned requirements. On the other hand, improper sampling may lead to decisive decisions on food safety and quality which could harm the consumer. In sampling for microbial analyses, the student or researcher should be aware of sample unit or size that should be withdrawn from the whole food or lot, randomization of sample withdrawing and conditions of sample handing after withdrawing until analyses completion. In research outcomes and in decision making on the safety, the fitness and quality of food, the accuracy and satisfaction of the sampling plan is more important than the number of determinations in a certain parameter.

Sample Size

For aerobic plate count, total coliforms, fecal coliforms, Escherichia coli (including enteropathogenic strains), *Staphylococcus* spp., *Vibrio* spp., *Shigella* spp., *Campylobacter* spp., *Yersinia* spp., *Bacillus cereus*, and Clostridium perfringens enumeration, 10 samples from each lot should be collected. Each sample should weigh about 226 g. The sample number and size for *Salmonella* spp. however, depends on the food category. In Food Category I which includes foods that would not normally be subjected to a process lethal to *Salmonella* spp. and are consumed by the aged, the infirmed, and infants, 60 samples should be collected from a lot. In Food Category II which includes foods that would not normally be subjected to a process lethal to *Salmonella* spp. between the time of sampling and consumption, 30 samples should be collected from a lot. In Food Category III which includes foods that would normally be subjected to a process lethal to *Salmonella* spp. between the time of sampling and consumption, 15 samples should be collected from a lot. For all categories, each sample should weigh about 375 g (Wallace and Hammack 2003).

Microbial samples are randomly withdrawn from a whole food. Different techniques such as computer and statistical software-based can be used to randomize the sample to represent the whole food.

Materials

- Sterile sampling bag
- Sterile scoop
- Marker
- Sterile gloves
- Lab coat
- Hair net
- Cool box
- Ice pack
- Thermometer

Methods

The sampling procedure was modified from Food Protection Services, Environmental Health Services Div., Canada (2010).

- Wash your hand with a sanitizer and wear a lab coat, gloves and hair net.
- Aseptically and randomly, withdraw the sample from a food lot and place it into a sterile sample bag.
- Immediately record the temperature of the food and its sensorial properties.
- Label the sample with a certain code and collect the following sample history data:
  - Collection site, date and time.
  - Food code number.
  - Lot size and number.
  - Name of the food.
  - State of the food (fresh/frozen/dried/salted/fermented/etc.).
  - Sample approximate weight.
  - Date of production and expiration of food.
  - Food storage conditions.
  - Store/shop/market hygienic status, temperature and relative humidity.
  - Type of analyses required.
  - Name of person who collected the sample.
- Transfer the sample to the laboratory as quick as possible from the time of collection in the original storage condition of the food. For instance, dry or canned food samples should be kept at room temperature during delivery to the laboratory. For high risk foods (fresh meat, fresh chicken, fresh fish) however, keep the sample in a cool box surrounded by ice packs. Frozen food samples should be kept in a re-chilled container surrounded by ice packs and kept solidly frozen at all times.
- Analyze the sample immediately upon arrival to the laboratory. The sample can be kept in a refrigerator but at temperatures ranging from 0°C to 4°C. For microbial analyses, samples must be analyzed within 24 h of collection.

- For microbial analyses, samples must not be frozen.
- A frozen sample should be thawed completely and rapidly. The sample must be analyzed immediately after thawing to avoid any chance for microbial growth after thawing. Sample should be thawed at temperatures outside the danger zone (5°C −65°C). In general, a frozen sample can be thawed at 0°C to 4°C. It should be kept at this temperature until thawing is complete.

Chapter 2

ENUMERATION OF SPECIFIC MICROORGANISMS IN FOODS

INTRODUCTION

This chapter deals with the enumeration techniques of specific microbial groups which are found in certain types of foods and receive special interest due to their spoilage, pathogen and antimicrobial activity potential. These microbes are currently enumerated using general media in many microbial manuals which could reduce the efficiency of these media to recover these microbes. Unlike most bacteria, marine bacteria require specific sodium chloride content in the media. This chapter provides the appropriate sodium chloride concentration in a non-selective medium which ensures the recovery of most marine, slightly halophilic bacteria that are found in seafood. Gram-positive bacteria are weakly-growing bacteria that are overcome by fast-growing Gram-negative bacteria. Therefore, if they are targeted in non-selective media, their recovery is influenced by Gram-negative bacteria. Moreover, some selective medium such as mannitol salt agar are selective to specific Gram-positive bacteria such as Staphylococci and Micrococci. This handbook, for the first time introduces a selective medium that inhibits the growth of Gram-negative bacteria and provides the selectivity for most Gram-positive bacteria. Although, it is used to enumerate Gram-positive bacteria from marine sources in this chapter, this medium can be used to enumerate Gram-positive bacteria from any food. Yeasts and molds are routinely enumerated using media such as potato dextrose agar. Since some yeasts produce well-degrading enzymes that could inhibit the growth of molds, their enumeration in one medium favors the growth of yeasts only. This chapter provides a selective medium with some inhibitors that inhibit the growth of bacteria and yeasts and provides the selectivity for the molds. This medium can be used to enumerate molds from wide sources including foods. Hydrogen sulfide producing bacteria such as *Shewanella putrefaciens* is a typical food spoiler that is needed to enumerate muscled foods to evaluate their quality and storage stability. The enumeration technique and conditions for this bacterium is rarely provided in current microbial manuals. This chapter provides a useful medium for enumeration of this bacterium based on it its ability to produce sulfide compounds, the typical compounds which are produced mainly by this bacterium. This medium was first used by Gram et al. (1987) to evaluate the spoilage of some freshwater fish.

ENUMERATION OF TOTAL AEROBIC BACTERIA

Aims

- To familiarize with spread-plate technique.
- To familiarize with bacterial enumeration techniques.
- To determine the microbial quality of fresh foods.

Introduction

The spread-plate technique or surface colony count provides a way of distributing small amounts of bacterial suspension across the surface of an agar plate. The major benefits of this system are, reduction in the amount of work, saving time and protecting heat-sensitive microorganisms (mostly psychrophils and psychtrophs) that are injured by pouring agar at 45°C (pour-plate technique).

The muscle tissue of a healthy animal contains few, if any bacteria, but the cut and exposed surfaces are easily contaminated after slaughtering and during handling. Microorganisms can multiply readily on the cut surfaces, although the microbial count of the interior of the meat usually remains much lower. After the dressing procedure, the microbial load of the cut surfaces will be 103 to 105 cfu/g. If the meat is kept at a low temperature, psychrotrophs such as *Pseudomonas* spp., *Acinetobacter* spp. and *Moraxella* spp. dominate the flora on the surface and could increase microbial spoilage. Log 5 cfu/g is considered as the limit of bacterial count in good quality food (ICMSF 1986).

Marine fish harbor many bacteria such as *Micrococcus* spp., *Bacillus* spp., *Staphylococcus* spp., *Streptococcus* spp., *Pseudomonas* spp., *Moraxella* spp., and *Flavobacterium* spp., The dominance of these genera depends on the fish's environment (marine or freshwater, tropical or cold waters), fish habitat (sea surface, bottom of the sea or coastal area) and fish body parts. For instance, fish from tropical and sub-tropical areas harbor more Gram-positive bacteria such as *Staphylococcus* spp, and *Streptococcus* spp. than Gram-negative bacteria. Moreover, the cultivation conditions such as the type of medium and incubation temperature lead to the dominance of certain bacteria on fish. In non-selective media such as standard plate count agar, the short-generation genera such as *Pseudomonas* spp. usually dominate the growth.

The bacteria found in potable water are natural bacteria, soil-dwelling bacteria and contaminants that normally inhabit the intestines of humans and animals. Most normal bacteria are Gram-negative bacteria such as *Pseudomonas* spp., *Flavobacterium* spp., *Acinetobacter* spp. Many aquatic bacteria will not grow on standard plate count agar, therefore counts on this medium will often be only one-tenth of those on CPS medium. The soil-dwelling bacteria which are washed by rain into streams include *Bacillus* spp. and Enterobacter spp. Most of these bacteria have an optimum temperature for growth at 25°C. Contamination from fecal material raises the presence of bacteria derived from the intestines which include *Escherichia coli, Enterococcus faecalis, Clostridium perfringens* and possibly *Salmonella* spp., E. coli O157:H7, *Campylobacter* spp. and *Vibrio cholerae.*

Materials

- Fresh food samples
- Pre-prepared plates of standard plate count agar, SPCA. The medium is prepared according to the instructions of the manufacturer.
- 225 ml and 9 ml of MRD
- Scissors, knife, forceps
- Petri dish, 2
- Top loading balance
- Stomacher bag
- Stomacher
- Stop watch
- Tibs 1 ml, 10
- Tibs dispenser
- Glass spreader

- Incubators at 35°C
- 70% ethanol
- Bunsen flame
- Beaker 250 ml
- Tissue
- Vertex
- Colony counter
- Chopping board

Methods

Sampling

- Aseptically remove about 50g of sample and cut into small pieces.
- Put a stomacher bag in the beaker and weigh 25g of the sample.
- Transfer 225 ml of MRD into the stomacher bag and blend for 1 min.

Diluting

Step 3 prepares dilution of 10^{-1}. Prepare further dilutions (based on the quality of foods e.g., good quality needs low dilutions, low quality needs high dilutions) by transferring 1 ml of the previous dilution into 9 ml MRD. For example to prepare dilution of 10^{-2}, transfer 1 ml of 10^{-1} to 9 ml MRD.

Plating

- Select 3 appropriate dilutions to be plated.
- Transfer 0.1 ml from each dilution into 2 plates of SPCA.
- Using the glass spreader, spread the samples over the plates.

1. Label 2 plates of SPCA as control.

Incubating

Incubate standard plate count agar plates inverted for 48 ± 2 h at 35°C (USFDA 2001).

Counting/Enumerating

Count all plates and record the data in the following table:

Sample	Dilutions									
	10^{-1}		10^{-2}		10^{-3}		10^{-4}		10^{-5}	
	Plates		Plates		Plate		Plates		Plates	
	1	2	1	2	1	2	1	2	1	2

Reporting the Bacterial Count

- Consider the plates with 30- 300 colonies for reporting the bacterial count in the sample.
- Report bacterial count as log cfu/g.
- Compare your bacteria count with log 5 cfu/g and evaluate the microbial quality of your sample.

ENUMERATION OF MARINE BACTERIA

Aim

To familiarize with marine bacterial growth and enumeration techniques.

Introduction

Marine bacteria differ from land or other foods' bacteria in their salt growth requirements. For instance, the total aerobic bacteria of marine origin were found to require 1.16 –1.75% NaCl for their growth (Malle and Valle, 1998). Therefore, NaCl content is needed to be carefully considered in the enriched medium for marine bacteria. In fact, among the common media which have been used to enumerate marine bacteria, marine salt agar (MSA) which consists of tryptone soya agar with 2% NaCl (Buller, 2004) showed a higher ability to recover total and psychrophilic marine bacteria as compared with other media (Al Bulushi 2009).

Materials

- Marine fish/water
- Pre-prepared plates of MSA (2% (w/v) NaCl tryptone soya agar). The base medium is prepared according to the instructions of the manufacturer.
- 225 ml and 9 ml of Most Recovery Diluent (MRD)
- Scissors, knife, forceps
- Petri dish, 2
- Top loading balance
- Stomacher bag
- Stomacher
- Stop watch
- Tibs 1 ml, 10
- Tibs dispenser
- Glass spreader
- Incubators at 25°C
- 70% ethanol

- Bunsen flame
- Beaker 250 ml
- Disinfectant
- Tissue
- Vertex
- Colony counter
- Loop
- Chopping board

Methods

Sampling

1. Aseptically remove about 50g of sample and cut into small pieces.
2. Put a stomacher bag in the beaker and 25g of the sample.
3. Transfer 225 ml of MRD into the stomacher bag and blend for 1 min.

Diluting

Step 3 in A prepares dilution of 10^{-1}. Prepare further dilutions by transferring 1 ml of the previous dilution into 9 ml MRD. For example to prepare dilution of 10^{-2}, transfer 1 ml of 10^{-1} to 9 ml MRD.

Plating

- Select 3 appropriate dilutions to be plated.
- Transfer 0.1 ml from each dilution into 2 plates of MSA.
- Using the glass spreader, spread the samples over the plates.
- Label 2 plates of MSA as control.

Incubating

Incubate plates of MSA inverted at 25°C for 72 hr.

Counting/Enumerating

Count all plates and record the data in the following table:

Sample	Dilutions									
	10^{-1}		10^{-2}		10^{-3}		10^{-4}		10^{-5}	
	Plates		Plates		Plate		Plates		Plates	
	1	2	1	2	1	2	1	2	1	2

Reporting the Bacterial Count

- Consider the plates with 30- 300 colonies for reporting marine total bacterial count in the sample.
- Report marine total bacterial count as log cfu/g.

Enumeration of Gram-Positive Bacteria

Aims

1. To understand Gram-positive bacteria and their importance.
2. To know the conditions and requirements of Gram-positive bacteria enumeration in foods.

Introduction

Gram-positive bacteria are widely found in different foods especially aquatic and mishandled foods. Most toxin-producers and many pathogens are in fact Gram-positive bacteria such as *Staphylococcus aureus, Clostridium botulinum, Bacillus cereus* and *Listeria monocytogenes*. As Gram-positive bacteria are weak competitors and slow-growing bacteria, they are overgrown by fast-growing Gram-negative bacteria.

Principle

Therefore, their enumeration even in selective media such as mannitol salt agar is a challenge for students and researchers. Due to its inhibitory effect on Gram- negative bacteria by breaking down the cell membrane of Gram-negative and inhibiting of DNA synthesis, phenylethyl alcohol in tryptone soya agar (PEA-TSA) was used successfully by Al Bulushi et al. (2008) to enumerate Gram-positive bacteria in sub-tropical marine fish.

Materials

- Food sample
- Pre-prepared plates of 0.25% (v/v) PEA-TSA (can be prepared by mixing 0.25 ml of phenylethyl alcohol in 100 ml tryptone soya gar). The base medium is prepared according to the instructions of the manufacturer.
- 225 ml and 9 ml of MRD
- Scissors, knife, forceps
- Petri dish, 2
- Top loading balance
- Stomacher bag
- Stomacher

- Stop watch
- Tibs 1 ml, 10
- Tibs dispenser
- Glass spreader
- Incubators at 32°C
- 70% ethanol
- Bunsen flame
- Beaker 250 ml
- Disinfectant
- Tissue
- Vertex
- Colony counter
- Loop
- Chopping board

Methods

Sampling

- Aseptically remove about 50g of sample and cut into small pieces.
- Put a stomacher bag in the beaker and 25g of the sample.
- Transfer 225 ml of MRD into the stomacher bag and blend for 1 min.

Diluting

Step 3 in A prepares dilution of 10^{-1}. Prepare further dilutions by transferring 1 ml of the previous dilution into 9 ml MRD. For example to prepare dilution of 10^{-2}, transfer 1 ml of 10^{-1} to 9 ml MRD.

Plating

- Select 3 appropriate dilutions to be plated.
- Transfer 0.1 ml from each dilution into 2 plates of 0.25% (v/v) PEA-TSA.
- Using the glass spreader, spread the samples over the plates.
- Label 2 plates of 0.25% (v/v) PEA-TSA as control.

Incubating

Incubate plates of 0.25% (v/v) PEA-TSA inverted at 32°C for 3 days.

Counting/Enumerating

Count all plates and record the data in the following table:

Sample	Dilutions									
	10^{-1}		10^{-2}		10^{-3}		10^{-4}		10^{-5}	
	Plates		Plates		Plate		Plates		Plates	
	1	2	1	2	1	2	1	2	1	2

Reporting the Bacterial Count

- Consider the plates with 30- 300 colonies for reporting Gram-positive bacterial count in the sample.
- Report Gram-positive bacterial count as log cfu/g.

ENUMERATION OF LACTIC ACID BACTERIA

Aims

1. To familiarize with anaerobic incubation techniques.
2. To familiarize with lactic acid bacterial enumeration techniques.

Introduction

Lactic acid bacteria (LAB) are a large group of useful bacteria used widely in food processing due to their ability to ferment certain carbohydrates and in food preservation due to their ability to produce antimicrobial agents such as bacteriocins. This group includes many genera such as *Lactobacillus, Lactococcus, Streptococcus, Enterococcus* and *Carnobacterium.* Lactic acid bacteria are Gram-positive rod and cocci bacteria, they are weakly proteolytic and lipolytic. Most LAB are mesophilic in general, they grow in the pH ranges from 4.0-4.5. Lactic acid bacteria are catalase and oxidase negative and facultative anaerobic, growing best in low concentrations of oxygen. They are isolated from foods to study their characteristics and antimicrobial activities.

Principle

Lactic acid bacteria are enumerated in MRS (ISO) agar (DE MAN, Rogosa and Sharpe). This medium is acidic with pH 5.7 which does not support the growth of many bacteria. Besides, MRS contains ammonium citrate and sodium acetate which inhibit the growth of contaminants such as molds and even certain LAB genera such as *Streptococcus*.

Materials

- Dairy products such as cheese
- Pre-prepared MRS plates. This medium is prepared according to the instructions of the manufacturer.
- Anaerogen

- Anaerobic jar
- 225 ml and 9 ml of MRD
- Scissors, knife, forceps
- Petri dish, 2
- Top loading balance
- Stomacher bag
- Stomacher
- Stop watch
- Tibs 1 ml, 10
- Tibs dispenser
- Glass spreader
- Incubator at 30°C
- 70% ethanol
- Bunsen flame
- Beaker 250 ml
- Tissue
- Vertex
- Colony counter
- Chopping board

Methods

Enumeration of Lactic Acid Bacteria

Sampling

- Aseptically remove about 50g of sample and cut into small pieces.
- Put a stomacher bag in the beaker and weigh 25g of the sample.

- Transfer 225 ml of MRD into the stomacher bag and blend for 1 min.

Diluting

Step 3 prepares dilution of 10^{-1}. Prepare further dilutions (based on the quality of foods e.g., good quality needs low dilutions, low quality needs high dilutions) by transferring 1 ml of the previous dilution into 9 ml MRD. For example to prepare dilution of 10^{-2}, transfer 1 ml of 10^{-1} to 9 ml MRD.

Plating

- Select 3 appropriate dilutions to be plated.
- Transfer 0.1 ml from each dilution into 2 plates of MRS.
- Using the glass spreader, spread the samples over the plates.
- Label 2 plates of MRS as control.
- Put the plates in the anaerobic jar containing anaerogen to provide a microaerophilic atmosphere (5% CO2).

Incubating

Incubate MRS plates inverted for 72 ± 2 h at 30°C (ISO 15214:1998).

Counting/ Enumerating

Count all plates and record the data in the following table:

Sample	**Dilutions**									
	10^{-1}		10^{-2}		10^{-3}		10^{-4}		10^{-5}	
	Plates		Plates		Plate		Plates		Plates	
	1	2	1	2	1	2	1	2	1	2

Reporting the Lactic Acid Bacterial Count

1. Consider the plates with 30- 300 colonies for reporting the bacterial count in the sample.
2. Report LAB count as log cfu/g.

Confirmation of Lactic Acid Bacteria

Conduct the following tests to confirm LAB.

Gram Sating

Lactic acid bacteria are Gram-positive.

Microscopic Test

Lactic acid bacteria are non-sporulated bacilli.

Catalase Test

Lactic acid bacteria are catalase negative.

Oxidase Test

Lactic acid bacteria are oxidase negative.

ENUMERATION OF YEASTS AND MOLDS IN FOODS

Aim

To familiarize with yeast and mold enumeration techniques in foods.

Introduction

Dried foods and high sugar foods are characterized with low water activity and high sugar content. These two elements provide the optimum growth requirements for many types of yeast and mold. Therefore, most foods associated with yeast and mold are dried food and fresh fruits such as grapes and citrus fruits. In the former, yeast and mold ferment carbohydrates to acids which reduce the pH and provide another growth requirement for these microorganisms, the pH. Yeast and mold growth in foods result in food spoilage which can be detected organoleptically by changing some of the food quality properties such as colour and flavor. Mold growth in foods results in food poisoning and toxicology due to production of mycotoxins such as aflatoxins and ochratoxins by certain strains of *Aspergillus*.

Principle

Yeast and mold together can be enumerated in non-selective media such as potato dextrose agar (PDA). As it is not a non-selective medium, it can provide the growth requirement for bacteria as well. Therefore, to inhibit bacterial growth in PDA, the medium is acidified to pH 3.5 by some organic acids such as lactic and tartaric acids.

Materials

- Dried foods
- Pre-prepared potato dextrose agar. The medium is prepared according to the instructions of the manufacturer.
- 10% lactic acid
- 225 ml and 9 ml of MRD
- Scissors, knife, forceps
- Petri dish, 2
- Top loading balance
- Stomacher bag

- Stomacher
- Stop watch
- Tibs 1 ml, 10
- Tibs dispenser
- Glass spreader
- Incubators at 25°C
- 70% ethanol
- Bunsen flame
- Beaker 250 ml
- Tissue
- Vertex
- Colony counter
- Chopping board

Methods

Sampling

- Aseptically remove about 50g of sample and cut into small pieces.
- Put a stomacher bag in the beaker and weigh 25g of the sample.
- Transfer 225 ml of MRD into the stomacher bag and blend for 1 min.

Diluting

Step 3 prepares dilution of 10^{-1}. Prepare further dilutions (based on the quality of foods e.g., good quality needs low dilutions, low quality needs high dilutions) by transferring 1 ml of the previous dilution into 9 ml MRD. For example, to prepare dilution of 10^{-2}, transfer 1 ml of 10^{-1} to 9 ml MRD.

Plating

- Select 3 appropriate dilutions to be plated.
- Transfer 0.1 ml from each dilution into 2 plates of PDA.
- Using the glass spreader, spread the samples over the plates.
- Label 2 plates of PDA as control.

Incubating

Incubate PDA inverted for 120 ± 2 h at 25°C (USFDA 2001).

Counting/Enumerating

Count only yeast and mold colonies. Bacterial colonies can be easily avoided by their morphology as being smaller in size compared with that of yeast and mold. Compile the data in the following table:

Sample	Dilutions									
	10^{-1}		10^{-2}		10^{-3}		10^{-4}		10^{-5}	
	Plates		Plates		Plate		Plates		Plates	
	1	2	1	2	1	2	1	2	1	2

Reporting the Yeasts and Molds Count

Report yeasts and molds count as log cfu/g.

ENUMERATION OF MOLD IN FOODS

Aim

To familiarize with mold enumeration techniques in foods.

Introduction

Mold grows in mediums which could provide a control mean for bacteria such as low pH, low water activity and high osmotic pressure media. Some of the mycotoxins which are produced by mold are carcinogenic or mutagenic and cause organ-specific pathology such as hepatotoxic (liver toxicity). On the other hand, some beneficial mold can be used in the food industry to produce specific flavor in foods such as their application in cheese. Moreover, certain types of mold such as *Rhizopus* spp. can spoil many fruits and vegetables.

Principle

Mold can be enumerated using dichloran-glycerol (DG18) agar base supplemented with chloramphenicol (Oxoid, UK). This medium contains glycerol which reduces the water activity of the medium and provides the optimum growth requirement for mold especially xerophilic mold and chloramphenicol which increases the selectivity of the medium for mold.

Materials

- Dried foods
- Prepared plates of dichloran-glycerol (DG18) agar base supplemented with chloramphenicol.
- The medium is prepared according to the instructions of the manufacturer.
- Chloramphenicol Supplement

- Glycerol
- 225 ml and 9 ml of MRD
- Scissors, knife, forceps
- Petri dish, 2
- Top loading balance
- Stomacher bag
- Stomacher
- Stop watch
- Tibs 1 ml, 10
- Tibs dispenser
- Glass spreader
- Incubators at 25°C
- 70% ethanol
- Bunsen flame
- Beaker 250 ml
- Tissue
- Vertex
- Colony counter
- Chopping board

Methods

Sampling

- Aseptically remove about 50g of sample and cut into small pieces.
- Put a stomacher bag in the beaker and weigh 25g of the sample.
- Transfer 225 ml of MRD into the stomacher bag and blend for 1 min.

Diluting

Step 3 prepares dilution of 10^{-1}. Prepare further dilutions (based on the quality of foods e.g., good quality needs low dilutions, low quality needs high dilutions) by transferring 1 ml of the previous dilution into 9 ml MRD. For example, to prepare dilution of 10^{-2}, transfer 1 ml of 10^{-1} to 9 ml MRD.

Plating

- Select 3 appropriate dilutions to be plated.
- Transfer 0.1 ml from each dilution into 2 plates of dichloran-glycerol (DG18) agar.
- Using the glass spreader, spread the samples over the plates.
- Label 2 plates of dichloran-glycerol (DG18) agar as control.

Incubating

Incubate dichloran-glycerol (DG18) agar inverted for 120 ± 2 h at 25°C.

Counting/Enumerating

Count only mold colonies. Compile the data in the following table:

Sample	Dilutions									
	10^{-1}		10^{-2}		10^{-3}		10^{-4}		10^{-5}	
	Plates		Plates		Plate		Plates		Plates	
	1	2	1	2	1	2	1	2	1	2

Reporting the Mold Count

Report mold counts as log cfu/g.

Enumeration of Hydrogen Sulfide Producing Bacteria in Foods

Aims

1. To study and characterize the microbial spoilage in foods.
2. To enumerate and characterize *Shewanella putrefaciens* as a hydrogen sulfide producing bacterium and study its spoilage potential.

Introduction

Organoleptically, sulfurous odor is one of the typical parameters used to evaluate the spoilage of muscled foods such as fish, meat and chicken. Shewanella spp. are typical spoilage bacteria in foods due their abilities to decompose amino acids and reduce trimethylamine oxide (TMAO). In fish, these bacteria decompose sulfur-containing amino acids such as cysteine and methionine to volatile sulfur compounds such as hydrogen sulfide (H2S) and methylmercaptan. These products produce the offensive sulfurous odors which make fish unfit for consumption. Moreover, *Shewanella* spp. reduces TMAO to trimethylamine (TMA) (Stenberg et al. 1984; Ringo et al. 1984, cited by Malle, et al. 1998) which is responsible for fishy odor in fish. As a result of their spoilage actions in fish, fish shelf-life can be predicted by estimating the level of H_2S producers and TMAO reducers. The main species of this genus which have been found to have spoilage potential are Sh. *putrefaciens* and Sh. *Baltica* (Vogel et al. 2005). These species are Gram-negative, motile rods, produce H2S and reduce TMAO.

Principle

Hydrogen sulfide producers such as Sh. *putrefaciens* can be enumerated in iron agar medium (Gram et al. 1987). In this medium, bacteria decomposes cysteine and forms black colonies due to the precipitation of iron sulfide (FeS). After spreading of bacterial suspension, the plates are overlaid with iron agar medium to prevent fading of the black colonies due to oxidation of FeS.

Materials

- 24–48 hour–old isolates
- Pre-prepared plates of iron agar (Gram et al. 1987):
- 2% peptone
- 0.3% lab lemco powder
- 0.3% yeast extract
- 0.03% ferric citrate
- 0.04% L-cysteine
- 0.5% sodium chloride
- 1.5% agar
- pH: 7.4 with 2N NaOH
- TMAO-medium (Gram et al. 1987):
- 2% peptone
- 0.3% Lab lemco powder
- 0.3% yeast extract
- 0.4% NaCl
- 0.4% KH_2PO_4
- 0.575% K_2HPO_4
- 0.05% $MgSO_4$
- 0.5% trimethylamine N-oxide dihydrate

- 0.2% agar
- pH: 6.8 with 2N NaOH
- Pre-prepared medium of sulfide indole motile medium, SIM medium. The medium is prepared according to the instructions of the manufacturer.
- Pre-prepared plates of 2% agar tryptone soya agar. The base medium is prepared according to the instructions of the manufacturer.
- Food sample, fish/meat/chicken
- 225 ml, 9 ml of MRD
- Stomacher
- Stomacher bag
- Scissors, knife, forceps
- Petri dish, 2
- Top loading balance
- Stop watch
- Tibs 1 ml, 10
- Tibs dispenser
- Glass spreader
- Incubator at 25°C
- 70% ethanol, 50 ml
- Bunsen flame
- Beaker 250 ml
- Tissue
- Vertex
- Colony counter
- Harrison's disc template
- Loop
- Chopping board
- Mineral oil
- Sterile needle

Methods

Preparation of Iron Agar Medium

- Mix all ingredients in water.
- Adjust the pH to 7.4.
- Autoclave at 115°C for 10 min.
- Cool to about 50°C and pour in plates.

Preparation of TMAO Medium

- Mix all ingredients in water.
- Adjust the pH to 6.8.
- Dissolve two tablets of resazurin per 500 ml of medium.
- Bring the medium to boiling until the tablets are dissolved.
- Dispense the medium into 8-9 ml test tubes.
- Autoclaved at 115°C for 10 min.

Enumeration of Sh. Putrefaciens

Sampling

- Aseptically remove about 50g of sample and cut into small pieces.
- Put a stomacher bag in the beaker and weigh 25g of the sample.
- Transfer 225 ml of diluent into the stomacher bag and blend for 1 min.

Diluting

Step 3 prepares dilution of 10^{-1}. Prepare further dilutions (based on the quality of foods e.g., good quality food needs low dilutions, low quality foods needs high dilutions) by transferring 1 ml of the previous dilution into 9 ml. For example, to prepare dilution of 10^{-2}, transfer 1 ml of 10^{-1} to 9 ml dilution.

Plating

- Select 3 appropriate dilutions to be plated.
- Transfer 0.1 ml from each dilution into 2 iron agar plates.
- Using the glass spreader, spread the samples over the plates.
- Label 2 plates of iron agar as control.

Incubating

Incubate iron agar plates inverted for 72 ± 2 h at 25°C.

Counting/Enumerating

Count only black colonies. Insert the colonies in the following table:

Sample	Dilutions									
	10^{-1}		10^{-2}		10^{-3}		10^{-4}		10^{-5}	
	Plates		Plates		Plate		Plates		Plates	
	1	2	1	2	1	2	1	2	1	2

Reporting the Bacterial count

- Consider the plates with 30- 300 colonies for reporting of Sh. putrefaciens count in the sample.
- Report Sh. putrefaciens count as log cfu/g.

Confirmation of Sh. Putrefaciens

- Using the Harrison's disc template select 3-5 colonies randomly from a plate containing 10-100 colonies.

- Streak each colony on 2% agar tryptone soya agar.
- Incubate the plates at for 72 ± 2 h at 25°C.
- Confirm *Sh. Putrefaciens* by:

Production of H_2S in SIM Medium

- Using a sterile needle, pick 24–48 hour–old isolates.
- Stab SIM medium to within 1 cm of the bottom of the tube. Keep the needle in the same line as it is inserted while removing it from the medium.
- Incubate the medium 72 ± 2 h at 25°C.
- Observe the colour change in SIM medium.

Result

Black colour in SIM medium starting from the stabbing line: positive for H_2S production. No change in SIM medium colour: negative for H_2S production

Reduction of TMAO to TMA

- Inoculate 1 ml aliquot of 24–48 hour–old isolates into TMAO medium.
- Immediately seal the medium with mineral oil.
- Incubate the medium 72 ± 2 h at 25°C.
- Observe the colour change in TMAO medium.

Result

Changing the colour from red to yellow: positive for TMAO reduction. No change in the TMAO medium: negative for TMAO reduction

Identification of Sh. Putrefaciens by Vitek 2 Compact

Further confirmation of *Sh. putrefaciens* can be achieved by vitek 2 compact identification.

Chapter 3

ENUMERATION AND IDENTIFICATION OF PATHOGENS IN FOODS

INTRODUCTION

Pathogens' enumeration and identification are conducted using the local and international standards. In fact, these standards usually abbreviate the methodology or do not provide the base and some basic requirements in the tests, believing that it is known by the users. Indeed, many users especially students find some difficulties to understand or even use these standards. To overcome these challenges, this chapter attempts to simplify the procedures of the standardized methodologies of pathogens' enumeration and identification. Moreover, it provides the principles of each pathogen enumeration on selective media. Besides enumeration methods, the chapter provides confirmation tests for the pathogens as recommended by the standards and media manufacturers. This chapter provides the enumeration and identification techniques for toxin-producing pathogens, namely *Staphylococcus aureus, Bacillus cereus*, sanitary bacteria, namely *Enterobacteriaceae, Escherichia coli,*

Salmonella spp., seafood contaminants namely *Vibrio parahaemolyticus* and emerging pathogens namely *Listeria monocytogenes*. These methods are based on international standards such as ISO and USFDA. The enumeration and identification of these pathogens are required to ensure food safety in universities, colleges, institutions and food safety authority laboratories.

Enumeration and Confirmation of *Staphylococcus aureus* in Foods

Aims

- To evaluate the safety and pathogen levels in foods.
- To familiarize with enumeration and confirmation techniques of *staphylococcus aureus.*

Introduction

Staphylococcus aureus is considered as a poor competitor in complex microbial ecosystems. Psychrotrophic bacteria may overgrow *S. aureus* in foods stored at refrigeration temperatures. The ability of *S. aureus* to produce intoxication depends on the ability of the food concerned to support the growth of *S. aureus*. *Staphylococcus aureus* is inactivated by pasteurization and inhibited by lactic acid fermentation. In respect to lactic acid fermentation, substances such as lactic acid, hydrogen peroxide and bacteriocins have been found to inhibit S. aureus. In contrast, slow acid production by the starters could permit growth of staphylococci to an extent enough for significant toxin production. Staphylococcal toxin C was found to be produced at 4% NaCl and 10% NaCl but no toxin was produced at 12% NaCl (Jay et al.

2005). Staphylococcal toxins can cause gastroenteritis at doses estimated 100-200ng toxins produced by 10^6 -10^7 cells/g or ml (Ray and Bhunia 2014).

Principle

Staphylococcus aureus is enumerated in Baird-Parker agar (BPA) which usually provides good recovery of *S. aureus,* even injured or stressed (Stewart 2003). Baird-Parker agar is enriched with egg yolk and potassium tellurite which inhibits the growth of other microorganisms. Tellurite is reduced to tellurium which gives black colonies, while the enzyme produced by *S. aureus* hydrolyzes egg yolk to precipitate insoluble salts of fatty acids which produce an opaque zone in the area surrounding the colony. Despite its ability to give good recovery of *S. aureus,* in particular those which have been injured by heat and other stresses, Baird-Parker agar showed poor selectivity for S. aureus from meat, cheese, meat sandwich and raw milk where 73% of 84 isolates from BPA were not *S. aureus*. Thus to confirm *S. aureus* in Baird-Parker agar, colonies are subjected to a coagulase test. Coagulase is an enzyme produced by *S. aureus* that converts fibrinogen (plasma glycoprotein) to fibrin.

Materials

- Food samples
- Pre-prepared plates of BPA enriched with egg yolk and potassium tellurite. The medium is prepared according to the instructions of the manufacturer.
- Pre-prepared plates of 2% agar standard plate count agar. The base medium is prepared according to the instructions of the manufacturer.

- 225 ml, 9 ml of Most Recovery Diluent (MRD)
- Stomacher
- Stomacher bag
- Scissors, knife, forceps
- Petri dish
- Top loading balance
- Stop watch
- Tibs 1 ml
- Tibs dispenser
- Glass spreader
- Incubator at 37°C
- 70% ethanol
- Bunsen flame
- Beaker 250 ml
- Tissue
- Vertex
- Colony counter
- Harrison's disc template
- Loops
- Chopping board
- Staphytect plus system

Methods

Enumeration of Staphylococcus Aureus

Sampling

- Aseptically remove about 50g of sample and cut into small pieces.

- Put a stomacher bag in the beaker and weigh 25g of the sample.
- Transfer 225 ml of MRD into the stomacher bag and blend for 1 min.

Diluting

Step 3 prepares dilution of 10^{-1}. Prepare further dilutions (based on the handling of food e.g., good handled food needs low dilutions, mishandled foods needs high dilutions) by transferring 1 ml of the previous dilution into 9 ml MRD. For example, to prepare dilution of 10^{-2}, transfer 1 ml of 10^{-1} to 9 ml MRD.

Plating

- Select 3 appropriate dilutions to be plated.
- Transfer 0.1 ml from each dilution into 2 plates of BPA.
- Using the glass spreader, spread the samples over the plates.
- Label 2 plates of BPA as control.

Incubating

Incubate BPA plates inverted for 48 ± 2 h at 37°C (USFDA 2001).

Counting/Enumerating

Count only black convex and shiny colonies with narrow white margins and are surrounded by clear zones. Insert the colonies in the following table:

Sample	**Dilutions**									
	10^{-1}		10^{-2}		10^{-3}		10^{-4}		10^{-5}	
	Plates		Plates		Plate		Plates		Plates	
	1	2	1	2	1	2	1	2	1	2

Reporting the bacterial count

- Consider the plates with 30- 300 colonies for reporting of *S. aureus* count in the sample.
- Report *S. aureus* count as log cfu/g.

Confirmation of Staphylococcus aureus

- Using the Harrison's disc template, select 3-5 colonies randomly from a plate containing 10-100 colonies.
- Streak each colony on a 2% agar standard plate count agar.
- Incubate the plates at for 48 ± 2 h at 37°C.
- Confirm *S. aureus* using Staphytect plus system which detects coagulase enzyme according to the manufacturer's instructions.

Result

Staphylococcus aureus is coagulase positive

ENUMERATION AND CONFIRMATION OF ENTEROBACTERIACEAE IN FOODS

Aims

- To evaluate the safety and pathogen levels in foods.
- To familiarize with enumeration and confirmation techniques of Enterobacteriaceae.
- To speciate Enterobacteriaceae by VITEK 2 compact.

Introduction

Enterobacteriaceae is a group of bacteria which includes large numbers of Gram-negative bacilli genera. Normally, these bacteria inhibit the intestines of humans and animals and the group includes many food pathogens such as *Escherichia* spp., *Shigella* spp., *Salmonella* spp., and biogenic-forming bacteria such as *Klebsiella* spp., *Hafnia* spp. and *Morganella* spp. Some pathogenic members produce enterotoxins which affect the intestinal tract, causing diarrhea and body fluid loss. These bacteria transmit to foods by inappropriate hand-washing after using the toilet or contamination with animals' feces, thus their presence in food indicates unhygienic food handling and contamination with animal sources.

Principle

Violet red bile glucose (VRBG) agar is an ISO enumerating medium for Enterobacteriaceae in foods. This medium is supplemented with crystal violet and bile salts which inhibit the growth of Gram-positive bacteria and allow Gram-negative bacteria to grow. Enterobacteriaceae rapidly ferment glucose and reduce the pH of the medium. These colonies are usually surrounded by purple halos of precipitated bile salts.

Materials

- Food samples
- 250 ml VRBG agar. The medium is prepared according to the instructions of the manufacturer.

- Pre-prepared plates of 2% agar standard plate count agar. The base medium is prepared according to the instructions of the manufacturer.
- Pre-prepared plates of nutrient agar. The medium is prepared according to the instructions of the manufacturer.
- Hugh and Leifson (1953) medium:
- 0.2% Peptone
- 0.5% NaCl
- 0.03% K_2HPO_4
- 0.2% agar
- 0.003% bromthymol blue
- 1% glucose
- pH 7.1
- 225 ml, 9 ml of Most Recovery Diluent (MRD)
- Stomacher
- Stomacher bag
- Scissors, knife, forceps
- Petri dish
- Top loading balance
- Stop watch
- Tibs 1 ml, 10
- Tibs dispenser
- Glass spreader
- Incubator at 37°C
- 70% ethanol, 50 ml
- Bunsen flame
- Beaker 250 ml
- Tissue
- Vertex
- Colony counter
- Harrison's disc template

- Loops
- Chopping board
- Oxidase test

Methods

Preparation of Hugh and Leifson (1953) Medium:

- Mix all ingredients *except glucose* in water and adjust the pH.
- Heat the medium until dissolving all ingredients.
- Dispense 5ml of the medium into 10 ml test tube.
- Autoclave at 121°C for 15 min. Cool the medium to about 45°C.
- Prepare 10% glucose solution. Sterilize glucose solution through membrane filtration (pore size of 0.45μm).
- Add 0.5ml of sterile glucose solution to 5ml of Hugh and Leifson (1953) medium. Mix by inverting, do not shake.

Enumeration of Enterobacteriaceae

Sampling

- Aseptically remove about 50g of sample and cut into small pieces.
- Put a stomacher bag in the beaker and weigh 25g of the sample.
- Transfer 225 ml of MRD into the stomacher bag and blend for 1 min.

Diluting

Step 3 prepares dilution of 10^{-1}. Prepare further dilutions (based on the handling of food e.g., good handled food needs low dilutions, mishandled foods needs high dilutions) by transferring 1 ml of the previous dilution into 9 ml MRD. For example, to prepare dilution of 10^{-2}, transfer 1 ml of 10^{-1} to 9 ml MRD.

Plating

- Select 3 appropriate dilutions to be plated.
- Transfer 1 ml from each dilution into 2 petri dishes.
- Pour 12-15 ml of VRBG agar (45-47°C) (pour plate technique).
- Leave the plate to solidify (≈ 15 min).
- Label 2 petri dishes as control and pour 12-15 ml of VRBG agar. Leave to solidify.

Incubating

Incubate VRBG agar plates inverted for 24 ± 2 h at 37°C (ISO 21528-2:2004).

Counting/Enumerating

Count only dark red to purple colored colonies larger than 0.5mm in diameter, usually surrounded by a purple zone. Insert the colonies in the following table:

Sample	**Dilutions**									
	10^{-1}		10^{-2}		10^{-3}		10^{-4}		10^{-5}	
	Plates		Plates		Plate		Plates		Plates	
	1	2	1	2	1	2	1	2	1	2

Reporting the Bacterial Count

1. Consider the plates with 30- 300 colonies for reporting of Enterobacteriaceae count in the sample.
2. Report Enterobacteriaceae count as log cfu/g.

Confirmation of Enterobacteriaceae

- Using the Harrison's disc template select 3-5 colonies randomly from a plate containing 10-100 colonies.
- Streak each colony on nutrient agar.
- Incubate the plates at for 24 ± 2 h at 37°C.
- Confirm Enterobacteriaceae by

Oxidase Test

- Touch the oxidase stick to the surface of colony thoroughly.
- Observe the color changes in oxidase stick.

Result

- Purple coloration within 5-10 sec: oxidase positive
- Absence of purple coloration: oxidase negative
- Enterobacteriaceae: oxidase negative

Oxidative/Fermentative Test

- Inoculate two test tubes of Hugh and Leifson (1953) medium with the bacterial growth.
- Seal one test tube with 0.3ml of sterile mineral oil.
- Incubate the medium at 25°C for 4 days.
- Observe the changes in the color.

Result

- Both tubes are green (no change): neutral, cannot use carbohydrates.
- Both tubes are yellow: fermentative.

- Open tube is yellow, sealed tube is green: oxidative
- Enterobacteriaceae: fermentative positive.

Vitek 2 Compact Identification

Identify Enterobacteriaceae to the genus and specie level by vitek 2 compact.

ENUMERATION AND CONFIRMATION OF *ESCHERICHIA COLI* IN FOODS

Aim

To familiarize with *Escherichia coli* enumeration and confirmation techniques.

Introduction

Escherichia coli is a Gram-negative, motile, nonspore-forming, rod-shape, facultative anaerobic bacterium. It is used as an index of possible fecal contamination and presence of enteric pathogens in foods. Enterotoxigenic *E. coli* (ETEC) strains cause diarrhea among travelers (also known as Traveler's Diarrhea). It is fatal in children due to excessive fluid loss and dehydration with a lethal dose of 10^8-10^9 cells (Ray and Bhunia 2014).

Principle

Escherichia coli is enumerated in tryptone bile X- glucuronide medium (TBX). This medium is supplemented with X-glucuronide, a substrate which detects glucuronidase activity, the enzyme that is highly produced by *Escherichia coli*. Glucuronidase utilizes X-glucuronide and releases the chromophore and the glucuronide. The released chromophore causes E. *coli* colonies to be colored blue/green.

Materials

- 250 ml of tryptone bile X- glucuronide medium (TBX). The medium is prepared according to the instructions of the manufacturer.
- Pre-prepared plates of 2% agar tryptone soya agar
- 225 ml, 9 ml of Most Recovery Diluent (MRD)
- Stomacher
- Stomacher bag
- Scissors, knife, forceps
- Petri dish,
- Top loading balance
- Stop watch
- Tibs 1 ml, 10
- Tibs dispenser
- Incubator at 35°C
- 70% ethanol
- Harrison's disc template
- Bunsen flame
- Beaker 250 ml
- Tissue
- Vertex
- Colony counter

Methods

Enumeration of Escherichia coli

Sampling

- Aseptically remove about 50g of sample and cut into small pieces.
- Put a stomacher bag in the beaker and weigh 25g of the sample.
- Transfer 225 ml of MRD into the stomacher bag and blend for 1 min.

Diluting

Step 3 prepares dilution of 10^{-1}. Prepare further dilutions (based on the handling of food e.g., good handled food needs low dilutions, mishandled foods needs high dilutions) by transferring 1 ml of the previous dilution into 9 ml MRD. For example, to prepare dilution of 10^{-2}, transfer 1 ml of 10^{-1} to 9 ml MRD.

Plating

- Select 3 appropriate dilutions to be plated.
- Transfer 1 ml from each dilution into 2 petri dishes.
- Pour 12-15 ml of TBX (45-47°C) (pour plate technique).
- Leave the plate to solidify (≈ 15 min).
- Label 2 petri dishes as control and pour 12-15 ml of TBX. Leave to solidify.

Incubating

Incubate TBX plates inverted for 24 ± 2 h at 35°C (USFDA 2002).

Counting/Enumerating

Count only blue/green colonies. Insert the colonies in the following table:

Sample	Dilutions									
	10^{-1}		10^{-2}		10^{-3}		10^{-4}		10^{-5}	
	Plates		Plates		Plate		Plates		Plates	
	1	2	1	2	1	2	1	2	1	2

Reporting the bacterial count

- Consider the plates with 30- 300 colonies for reporting of *E. coli* count in the sample.
- Report *Escherichia coli* count as log cfu/g.

Confirmation of Escherichia coli

- Using the Harrison's disc template select 3-5 colonies randomly from a plate containing10- 100 colonies.
- Streak each colony on 2% agar tryptone soya agar .
- Incubate the plates at for 24 ± 2 h at 35°C.
- Confirm Escherichia coli by vitek 2 compact identification

DETECTION AND CONFIRMATION OF *SALMONELLA* SPP. IN FOODS

Aims

- To familiarize with *Salmonella* spp. detection and confirmation techniques.

- To evaluate the safety of foods.

Introduction

Salmonella spp. are typical, active and heavily involved pathogens in food poisoning of Enterobacteriaceae. *Salmonella* spp. are Gram-negative motile bacteria. Generally, they produce hydrogen sulfide, decarboxylate lysine and ornithine. They grow optimally between 35°C to 37°C. *Salmonella* spp. are sensitive to pH 4.5 or below and viable at water activity 0.94.

Principle

Salmonella spp. are enumerated first by inoculating in Rappaport-Vassiliadis (RV) enrichment broth then spreading on xylose-lysine-desoxycholate agar (XLD) following incubation (Sebastia et al. 2012). In Rappaport-Vassiliadis (RV) enrichment broth, the combination actions of malachite green which provides inhibitory action for other bacteria excluding *Salmonella* spp., the high magnesium chloride content which increases the osmotic pressure, and low pH of the broth make this broth selective for *Salmonella* spp. In XLD agar, *Salmonellae* utilize xylose and change the pH of the medium to alkaline and decarboxylate.

Xylose-lysine-desoxycholate agar is a selective medium for the isolation of *Salmonella* spp. and *Shigella* spp. *Salmonella* spp. can be presumptively differentiated from *Shigella* spp by a hydrogen sulfide test where *Salmonella* spp. can produce hydrogen sulfide whereas *Shigella* spp. cannot. *Salmonella* spp. also share the biochemical characteristics with closely genus *Edwardsiella* spp. Presumptively, *Salmonella* spp. can be differentiated from *Edwardsiella* spp. by citrate utilization and indole production tests, where *Salmonella* spp are positive for citrate test and negative for indole test, whereas, *Edwardsiella* spp are negative for citrate test and positive for indole test.

Materials

- Food sample, fish/ meat/ chicken
- Buffered peptone water (BPW), 225 ml
- Rappaport-Vassiliadis (RV) enrichment broth, 3 test tubes of 10 ml
- Pre-prepared plates xylose-lysine-desoxycholate agar (XLD), 6 plates
- Hydrogen sulfide test set
- Citrate utilization test set
- Indole production test set
- Stomacher
- Stomacher bag
- Scissors, knife, forceps
- Petri dish, 2
- Top loading balance
- Stop watch
- Tibs 1 ml, 10
- Tibs dispenser
- Glass spreader
- Incubators at 37°C, 42°C
- 70% ethanol, 50 ml
- Bunsen flame
- Beaker 250 ml
- Tissue
- Vertex

- Loop
- Chopping board

Methods

Media Preparation

All media are prepared according to manufacturer instructions.

Salmonella spp. Detection

Step 1

- Prepare a homogenous sample by cutting and chopping about 50 g of the sample.
- Mix 25 g of sample with 225 ml BPW.
- Blind the mixture in stomacher for 1 min.

Step 2

- Inoculate 3 test tubes of 10 ml Rappaport-Vassiliadis (RV) enrichment broth each with 0.1 ml of BPW dilution (step 1).
- Incubate the inoculated broth at 42°C for 48 hr.

Step 3

- Streak 0.1 ml of inoculated Rappaport-Vassiliadis (RV) enrichment broth (step 2) into 3 plates of XLD agar, each with 0.1 ml.
- Incubate the plates at 37°C for 24 hours.
- Consider only red colonies as a presumptive detection of *Salmonella* spp.

Salmonella spp. Confirmation

The suspected *Salmonella* spp. should be purified and confirmed as following:

H_2S Test

To presumptively differentiate *Salmonella* spp. from *Shigella* spp.

Result

- *Salmonella* spp.: positive
- *Shigella* spp.: negative.

Citrate Utilization Test

To presumptively differentiate *Salmonella* spp. from *Edwardsiella* spp.

Result

- *Salmonella* spp.: positive
- *Edwardsiella* spp.: negative.

Indole Production Test

To presumptively differentiate *Salmonella* spp. from *Edwardsiella* spp.

Result

- *Salmonella* spp.: negative
- *Edwardsiella* spp.: positive.

Further molecular or phenotypic identification is needed to confirm *Salmonella* spp. detection.

ENUMERATION AND CONFIRMATION OF *LISTERIA MONOCYTOGENES* IN FOODS

Aims

1. To familiarize with *Listeria monocytogenes* detection and confirmation techniques.
2. To evaluate the safety of foods.

Introduction

Listeria monocytogenes is a Gram-positive pathogen causing listeriosis, a long-time known food poisoning. Human listeriosis is highly fatal to fetuses, newborn, infants, the elderly, pregnant and immunocompromised people. *Listeria monocytogenes* is motile at temperature ranged from 20°C to 30°C and grows rapidly at 7°C to 10°C. It resists freezing, drying, > 10% salt and pH 5 and above, but sensitive to pasteurization temperatures.

Principle

Listeria monocytogenes is enumerated firstly by growing in Half-Fraser broth and secondly by spreading on Palcam agar (Sebastia et al. 2012). Both media contain aesculin and ferric ammonium citrate. *Listeria* spp. hydrolyze aesculin to aesculetin which reacts with ferric ions resulting in blackening of the broth and producing a black halo around colonies in Palcam agar. Changing the color of broth and producing black colonies can be considered as a presumptive indicator of *Listeria* spp. presence which are needed to be confirmed and differentiated by molecular and phenotypic techniques. *Listeria monocytogenes* media are heavily loaded with many inhibitors such as of lithium chloride, ceftazidime, polymyxin B and acriflavine hydrochloride to maximize the media selectivity for *Listeria monocytogenes*.

Materials

- Food sample
- Half-Fraser broth base
- Half-Fraser broth supplement
- Palcam agar base
- Palcam selective supplement
- Stomacher
- Stomacher bag
- Scissors, knife, forceps
- Petri dish, 2
- Top loading balance
- Stop watch
- Tibs 1 ml, 10
- Tibs dispenser
- Glass spreader
- Incubators at 30°C, 37°C
- 70% ethanol, 50 ml
- Bunsen flame
- Beaker 250 ml
- Tissue
- Vertex

- Loop
- Chopping board

Methods

Media Preparation

All media are prepared according to manufacturer instructions.

Listeria monocytogenes Enumeration

Step 1

- Prepare a homogenous sample by cutting and chopping about 50 g of the sample.
- Mix 25 g of sample with 225 ml of Half-Fraser broth.
- Blend the mixture in stomacher for 1 min.
- Incubate the blended sample at 30°C for 48 h.
- Check the presumptive presence of *Listeria monocytogenes* by turning the color of Half-Fraser broth to black.

Step 2

- Streak 0.1 ml of black-colored Half-Fraser broth (step 1) and a 0.1 ml of yellow-colored Half-Fraser broth into Palcam medium plates.
- Incubate the plates at 37°C for 24 hours.

Result

- Grey-green colony with a black sunken center and a black halo: suspected *Listeria monocytogenes*.

- Yellow Colony: no *Listeria monocytogenes.*

Confirmation of Listeria monocytogenes

The suspected *Listeria monocytogenes* should be purified and confirmed by molecular techniques and phenotypic techniques such as API *Listeria*.

ENUMERATION AND CONFIRMATION OF *VIBRIO PARAHAEMOLYTICUS* IN FOODS

Aims

- To familiarize with *Vibrio parahaemolyticus* enumeration and identification techniques.
- To evaluate the safety of seafood

Introduction

Vibrio parahaemolyticus is an aquatic pathogen that causes gastroenteritis. The food vehicles of this type of pathogen are coastal water harvested seafood such as oysters, shrimps, crabs, lobsters and related shellfish indicating high potential for cross-contamination from human waste. *Vibrio parahaemolyticus* is Gram-negative bacterium requiring 1-8% NaCl for growth. It grows optimally at temperatures between 30°C-35°C, pH from 7.6-8.6 and water activity of 0.992. It is a heat sensitive bacterium with $D_{47°C}$ values of 0.8-65.1 min.

Principle

Vibrio parahaemolyticus enumerated on thiosulfate citrate bile salts sucrose (TCBS) agar which is a selective medium for *Vibrio* spp.

Thiosulfate Citrate bile salts sucrose (TCBS) agar provides selectivity for *Vibrio* spp. through its typical composition. For instance, sodium thiosulphate, sodium citrate which provide alkaline pH besides Ox bile inhibit the growth of Gram-positive bacteria, while sodium chloride optimizes the growth and metabolic activity of halophilic *Vibrio* spp. Sucrose, the fermenting sugar is utilized by *Vibrio cholera* to produce yellow colonies by fermenting sucrose to acid which reduces the pH of the medium, and as a result, bromothymol blue, the indicator in TCBS is changed from blue to yellow. *Vibrio parahaemolyticus* which does not utilize sucrose, produces blue-green colonies (Rosec et al. 2012).

Materials

- Seafood (typically: oysters, shrimps, crabs)
- Pre-prepared plates of TCBS
- 1% NaCl nutrient agar
- 225 ml alkaline saline peptone water (ASPW) (typical composition: NaCl, 20g/l and peptone, 20g/l, the pH is adjusted to 8.6 with 5 N NaOH)
- 9 ml ASPW
- Stomacher
- Stomacher bag
- Scissors, knife, forceps
- Petri dish, 2
- Top loading balance
- Stop watch
- Tibs 1 ml, 10
- Tibs dispenser
- Glass spreader
- Incubators at 37°C, 42°C
- 70% ethanol, 50 ml

- Bunsen flame
- Beaker 250 ml
- Tissue
- Vertex
- Colony counter
- Harrison's disc template
- Loop
- Chopping board

Methods

Media Preparation

The medium and dilution solutions are prepared according to manufacturer instructions.

Vibrio parahaemolyticus Enumeration

Sampling

- Aseptically remove about 50g of sample and cut into small pieces.
- Put a stomacher bag in the beaker and weigh 25g of the sample.
- Transfer 225 ml of ASPW into the stomacher bag and blend for 1 min.
- Incubate the bacterial suspension at 42°C for 6 hr.

Diluting

Step 3 prepares dilution of 10^{-1}. Prepare further dilutions by transferring 1 ml of the previous dilution (step A) into 9 ml ASPW.

For example to prepare dilution of 10^{-2}, transfer 1 ml of 10^{-1} to 9 ml ASPW.

Plating

- Select 3 appropriate dilutions to be plated.
- Transfer 0.1 ml from each dilution into 2 TCBS platcs.
- Using the glass spreader, spread the samples over the plates.
- Label 2 plates of TCBS agar as control.

Incubating

Incubate TCBS agar plates inverted at 37°C for 24h.

Counting/ Enumerating

Count only blue-green colonies. Insert the colonies in the following table:

Sample	**Dilutions**									
	10^{-1}		10^{-2}		10^{-3}		10^{-4}		10^{-5}	
	Plates		Plates		Plate		Plates		Plates	
	1	2	1	2	1	2	1	2	1	2

Confirmation of Vibrio parahaemolyticus

- Using the Harrison's disc template select 3-5 colonies randomly from a plate containing 10- 100 colonies.
- Streak each colony on 1% NaCl nutrient agar.
- Incubate the plates at for 24 ± 2 h at 37°C.
- Confirm *Vibrio parahaemolyticus* by API 20E gallery.

ENUMERATION AND IDENTIFICATION OF *BACILLUS CEREUS* IN FOODS

Aims

1. To familiarize with *Bacillus cereus* enumeration and identification techniques.
2. To evaluate the safety of dried foods.

Introduction

Bacillus cereus is a group of pathogens including many members such as *Bac. cereus, Bac. anthracis* and *Bac. mycoides*. Certain members such as *Bac. anthracis* causes fatal diseases such as anthrax in humans and animals. The bacteria of this group are aerobic, gram-negative and motile at 30°C. *Bacillus cereus* grows optimally at 35°C-40°C. They survive at high temperatures used in cooking but are sensitive to pasteurization. These pathogens produce a wide range of toxins which are highly heat stable and some of these toxins can be found in certain foods such as rice and pasta.

Principle

Bacillus cereus can be enumerated in wide ranges of media such as red mannitol agar, mannitol egg yolk-polymyxin agar and *Bacillus cereus agar* (BCA). The former is recommended by ISO 7932:2004 (Sebastia et al. 2012). *Bacillus cereus* agar is supplemented with polymyxin B as a selective agent which inhibits the growth of other bacteria and egg yolk which is used to detect lecithinase which is present in most *Bacillus cereus* strains. *Bacillus cereus* strains utilize lecithin in egg yolk and produce whitish precipitated colonies. Besides the supplements, BCA contains mannitol and phenol red as an indicator which produces the red color in BCA. Since *Bacillus cereus* strains do not ferment mannitol, they produce pink colonies surrounded by a halo of precipitate indicating the production of lecithinase in BCA. Other bacteria ferment mannitol, and change the pH of BCA. As a result, they produce yellow colonies.

Materials

- Food samples: raw rice
- Pre-prepared plates of BCA supplemented with egg yolk and polymyxin B. The medium is prepared according to the instructions of the manufacturer.
- Compass *Bacillus cereus* agar. This medium is prepared according to the instructions of the manufacturer.
- 225 ml, 9 ml of buffered peptone water (BPW)
- Stomacher
- Stomacher bag
- Scissors, knife, forceps
- Petri dish
- Top loading balance
- Stop watch
- Tibs 1 ml
- Tibs dispenser
- Glass spreader
- Incubator at 30°C
- 70% ethanol
- Bunsen flame
- Beaker 250 ml
- Tissue

- Vertex
- Colony counter
- Harrison's disc template
- Loops
- Chopping board

Methods

Enumeration of Bacillus cereus

Sampling

- Aseptically remove about 50g of sample and cut into small pieces.
- Put a stomacher bag in the beaker and weigh 25g of the sample.
- Transfer 225 ml of BPW into the stomacher bag and blend for 1 min.

Diluting

Step 3 prepares dilution of 10^{-1}. Prepare further dilutions (based on the handling of food e.g., good handled food needs low dilutions, mishandled foods needs high dilutions) by transferring 1 ml of the previous dilution into 9 ml BPW. For example, to prepare dilution of 10^{-2}, transfer 1 ml of 10^{-1} to 9 ml BPW.

Plating

- Select 3 appropriate dilutions to be plated.
- Transfer 0.1 ml from each dilution into 2 plates of BCA.
- Using the glass spreader, spread the samples over the plates.
- Label 2 plates of BCA as control.

Incubating

Incubate BCA plates inverted for 24 ± 2 h at 30°C.

Counting/Enumerating

Count only pink color and irregular edge colonies surrounded by a white area. Insert the colonies in the following table:

Sample	Dilutions									
	10^{-1}		10^{-2}		10^{-3}		10^{-4}		10^{-5}	
	Plates		Plates		Plate		Plates		Plates	
	1	2	1	2	1	2	1	2	1	2

Reporting the Bacterial Count

- Consider the plates with 30- 300 colonies for reporting of *Bacillus cereus* count in the sample.
- Report *Bacillus cereus* count as log cfu/g.

Confirmation of Bacillus cereus

- Using the Harrison's disc template select 3-5 colonies randomly from a plate containing 10-100 colonies.
- Streak each colony on compass *Bacillus cereus* agar.
- Incubate the plates at for 24 ± 2 h at 30°C.

Result

- *Bacillus cereus* strains form green colonies in compass *Bacillus cereus* agar.
- Further confirmation can be achieved by using specific *Bacillus cereus* primers technique.

Chapter 4

MICROBIAL IDENTIFICATION TECHNIQUES

INTRODUCTION

Many techniques in microbial identification are developed continually to the extent that it might be difficult to cope with all new developed techniques. This chapter provides some of these techniques which have been used for a long time and their efficiencies evaluated. Vitek is one of most common techniques used by many laboratories. This technique initially has been used in the clinical areas where the database has been built-up to include common human pathogens. In later years, the manufacturer has attempted to expand this technique's database to include environmental microorganisms from foods and other sources. Besides this, the manufacturer is updating the exiting database to include the identification of new microbes in clinical and environmental areas. Most pathogens and food spoilers can be identified using this technique.

This chapter includes one version of vitek called vitek 2 compact. The system is simple in application so that any user with any background level can learn it within a few days. However, this system as with all commercial identification does not provide the identification for all microbes besides other factors, included in this chapter that affect the performance of this system. Molecular identification techniques overcome the limitation of the commercial identification systems in microbial identification. However, these techniques need extensive training and a lot of preparation. This chapter provides simple and easy follow-up procedures of molecular identification so that any user with limited training will be able to implement these techniques. Two of most commonly used molecular techniques, namely specific genus and specie primers and sequencing of 16S rRNA gene are included in this chapter. The former can be considered as the shortest molecular technique. This technique is used mainly to confirm the identification of suspected microbes in cases such as food poisoning. The latter technique is a common and general molecular technique that is used in identification of unknown microbes. Besides these techniques, this handbook provides the primers of most common pathogens which can be used in food poisoning to confirm the causes of microbial food poisoning.

Identification of Microorganisms by VITEK 2 Compact

Aims

- To understand the concept of phenotypic methods in bacterial identification.
- To introduce VITEK 2 compact as a commercial identification technique.

- To identify food bacteria by VITEK 2 compact.

Introduction

In commercial methods, many devices and techniques such as VITEK and API are used to identify microorganisms. These are based on the microbial ability to utilize substrates such as different carbohydrates and amino acids or show sensitivity and resistance to some antibiotics. Commercial methods provide fast and simple methods of identification, however, the databases do not include the chemical profiles of some microorganisms. As a result, these systems cannot provide identification for all microorganisms.

Principle

VITEK 2 compact is an automated identification system which uses growth-based technology for microbial identification. This technique provides identification for industrial microorganisms at low to mid-level for clinical microorganisms of Gram-negative bacteria, Gram-positive bacteria, Gram-positive spore-forming bacilli and yeast and yeast-like organisms. The system uses cards with 64 wells which contain substrates that measure acidification, alkalinization, enzyme hydrolysis and susceptibility to different antibiotics. VITEK identifies microorganisms on their abilities to metabolize the card substrates. This results in information on different color images which are analyzed and compared with those produced by certain microbes in the system databases. Each identification is provided with the probability level. Excellent identification provides 96-99% probability whereas, acceptable identification is 85-88% probability. Microorganisms which are identified by VITEK are needed to be pure, strong enough to use the substrates, and included in the system databases.

Materials

- Fresh young colonies of a bacterium
- 0.45% NaCl.
- 3% KOH
- Vitek cards
- Vitek system

Methods

Determination of Gram Reaction

- Place a clean slide on tissue paper.
- Put one drop of 3% KOH solution on the slide
- Pick off a little of the bacterial growth.
- Stir the bacterial growth in a circular motion in the KOH solution for 30 sec.
- Raise the loop slowly and observe the texture of bacterial growth in KOH solution.

Result

- Viscous texture with stringing occurred within 30 sec.: Gram-negative bacteria.
- Watery texture with absence of stringing: Gram-positive bacteria.

Note: the result should be considered within 30 sec of stirring. Repeat the test three times to confirm the results.

Bacterial Suspension Preparation

- A microorganism is grown on non-selective medium (preferable tryptone soya agar) for 24 hours at the optimum temperature of that microorganism.
- Bacterial colonies are mixed with 3ml 0.45% NaCl.
- The density of the suspension are checked and adjusted as follows:
  - Gram-positive and Gram-negative cards: 0.5-0.63.
  - Bacillus and yeast cards: 1.8-2.20.

Card Loading and Incubation

A card is prepared and loaded into VITEK 2 compact according to the instructions of the manufacturer. After incubation, the unknown bacteria are identified based on availability in the VITEK 2 compact database.

Identification of Microorganisms by Specific Genus and Species Primers

Aims

- To understand the principles of microbial identification by specific genus and species primers.
- To familiarize with the techniques of application of specific primers in microbial identification.

Introduction

Using specific primers to identify certain microbes is an essential and helpful technique in certain cases especially in cases such as food poisoning where the suspected pathogens need confirmation at a high level in a very short time to confirm the case and take the necessary steps. Fortunately, the primers of common pathogen species are known and produced commercially (included in this book).

Principle

The identification of microbes by genus and specie primers is based on comparison of the DNA of an unknown microbe with that of a known microbe in the same reaction. Practically, identification of microbes using specific primers involves preparing the DNA template of the suspected microbe, amplification, purification and comparison.

In amplification, the mixture of polymerase chain reaction (PCR) and DNA template is heated at high temperature which causes the denaturation of hydrogen bonds between DNA strands to release the DNA single strand. Immediately, the extension process starts where DNA polymerase begins to combine the DNA of the unknown microbe with the primers of the known microbe to produce an amplicon which can be defined as synthesized DNA.

The amplicon is purified and visualized by electrophoresis gel. The unknown microbe is identified by comparing the size of its amplicon with that of the known microbe. This technique does not need any sequencing of DNA.

Materials

- Primers of a certain strain of *Staphylococcus aureus* (an example, and more primers are included in this text book).
- Colonies of suspected *Staphylococcus aureus*
- Polymerase chain reaction components:
  - 10x PCR buffer containing KCl and Tris-HCl (pH 8.4)
  - 25 mM $MgCl_2$
  - 1.25 mM deoxyribonucleotide triphosphate (dNTP's)
  - 25 pmole/µl forward primer (sa442-1)
  - 25 pmole/µl reverse primer (sa442-2)
  - 10 mg/ml bovine serum albumin (BSA)
  - Taq
  - UV sterilized deionized water
- Low EEO agarose gel
- 1xTBE (Tris, Boric acid, EDTA) buffer pH 8.3
- 25 mg/ml ethidium bromide
- Thermal cycler

Methods

The PCR protocol varies from one laboratory to another. The following protocol is based on a protocol followed by Al Bulushi (2009) and Al Bulushi et al. (2010).

DNA Template Preparation

- Take a loop-full of fresh colonies of unknown *S. aureus* strain.

- Add UV sterilized deionized water (e.g., 200 µl).
- Mix well.

Polymerase Chain Reaction Protocol

- Mix PCR components.
- Add 11.8 µl UV sterilized deionized water.
- Add DNA template.
- Mix well.

Amplification

- Add the PCR mixture containing DNA template in the thermal cycler.
- Run the following amplification program:
- One cycle at 94°C for 5 min.
- 30 cycles at 94°C for 30 sec., 55°C for 30 sec., 72°C for 30 sec.
- Extension at 72°C for 7 min.

Purification and Visualization

- Apply the amplified products in 2% low EEO agarose gel.
- Dip the gel in 1X TBE.
- Run the gel at 100 v for 40 min.

Identification of S. aureus Strain

Measure the size of amplicon developed in the gel. Certain strains of *Staphylococcus aureus* provide an amplicon with the size of 108 bp.

IDENTIFICATION OF MICROORGANISMS BY SEQUENCING OF 16S RRNA GENE

Aims

1. To understand the concept of genotypic methods in bacterial identification.
2. To familiarize with the sequencing of 16S rRNA gene technique of bacterial identification.

Introduction

Genotypic and commercial methods are routinely used in microbial identification. However, the decision for selecting these techniques is usually based on many factors such as the level of accuracy, number of samples, time, cost, labor, database of the identification system, facilities, metabolic activity and the genus or specie.

Principle

Genotypic methods are based on identification the sequences of DNA and comparing these sequences with that of known microorganisms in the genbank. An A sequence can be defined as the arrangement of nucleotides (adenine (A), thymidine (T), cytosine (C), guanine (G). A always pairs with T and C always pairs with G. Example of a sequence:

TTAACGGGGCCCTTTAAA....... TTTAAACCCGGGTTT
AATTGCCCCGGGAAATTT...... AAATTTGGGCCCAAA

In commercial methods, many devices and techniques such as VITEK and API are used to identify microorganisms. These are based on the microbial ability to utilize substrates such as different carbohydrates and amino acids or show sensitivity and resistance to some antibiotics. Commercial methods provide fast and simple methods of identification, however, the databases of these techniques do not include the chemical profiles of some microorganisms. As a result, these systems cannot provide identifications for all microorganisms. Thus, in these cases, the identification of microbes by sequencing of 16S rRNA gene becomes the best option.

In the identification of microbes by the sequencing of 16S rRNA gene, the sequences of DNA of unknown microbes are compared with sequences of known microbes available in the genbank. Practically, identification of microbes by sequencing of 16S rRNA gene involves preparation of the DNA template of the suspected microbe, amplification, purification, cleaning up, sequencing and comparison.

In amplification, the mixture of polymerase chain reaction (PCR) and DNA template is heated at a high temperature which causes the denaturation of hydrogen bonds between DNA strands to release the DNA single strand. Immediately, the extension process starts where DNA polymerase begins to combine the DNA of an unknown microbe with the primers of a known microbe to produce an amplicon which can be defined as synthesized DNA. The amplicon is purified and visualized by electrophoresis gel. The amplicon is further cleaned up using a special kit and sent for sequencing in specialized centers. The unknown microbe is identified by comparing its sequences with that of known microbe in the genbank through a blast search.

Materials

- Universal forward primer

- Universal reverse primer
- Colonies of unknown microbe
- Polymerase chain reaction components:
  - 10x PCR buffer containing KCl and Tris-HCl (pH 8.4)
  - 25 mM $MgCl_2$
  - 1.25 mM deoxyribonucleotide triphosphate (dNTP's)
  - 25 pmole/µl forward primer
  - 25 pmole/µl reverse primer
  - 10 mg/ml bovine serum albumin (BSA)
  - Taq
  - UV sterilized deionized water
  - Low EEO agarose gel
  - 1xTBE (Tris, Boric acid, EDTA) buffer pH 8.3
  - 25 mg/ml ethidium bromide
  - Loading buffer
  - Thermal cycler
  - PCR clean-up system
  - Centrifuge

Methods

The PCR protocol varies from one laboratory to another. The following protocol is based on a protocol followed by Al Bulushi (2009) and Al Bulushi et al. (2010).

DNA Template Preparation

- Take a loop-full of fresh colonies of unknown *S. aureus*
- Add UV sterilized deionized water (e.g., 200 µl).
- Mix well.

Polymerase Chain Reaction Protocol

- Mix PCR components.
- Add 12.5 µl UV sterilized deionized water.
- Add DNA template.
- Mix well.

Amplification

- Add the PCR mixture containing DNA template in the thermal cycler.
- Run the following amplification program:
- One cycle at 94°C for 5 min.
- 30 cycles at 94°C for 30 sec., 55°C for 30 sec., 72°C for 60 sec.
- Extension at 72°C for 7 min.

Purification and Visualization

- Mix the amplified products with loading buffer
- Apply amplified products in 2% low EEO agarose gel stained with ethidium bromide.
- Dip the gel in 1X TBE.
- Run the gel at 100 v for 40 min.

Cleaning-up

Step 4 prepares DNA which needs to be cleaned up. A special DNA clean-up kit is available commercially and all instructions for clean-up are available on the label.

Sequencing and Blast Search

Before sequencing, the DNA is mixed with the forward and reverse primers. The DNA can be sent to any specialized center for sequencing. Microbial DNA can be sequenced either partially or completely. The sequences can be submitted to the genbank through a blast search which can compare the sequences of unknown microbes with those of known microbes and provide the closest possible similarity. In general, the microbial identify can be accepted with a similarity of 97% and above.

Example: the sequences of *Staphylococcus aureus strain* PVM 715 (NCBI, US)

AGATGCTGGTACAGGTATCCGTGAATACAACGATGGAAC ATTTGGATATGAAGCGAGACCAAGATTCAAC

AAGCCAAGTGAAACAAATGCATACAACGTAACGACAAAT CAAGATGGCACAGTAACATATGGCGCTCGCC

CAACACAAAATAAGCCATCAGAAACAAATGCATATAACG TAACAACACATGCAAACGGCCAAGTGTCATA

CGGTGCTCGTCCGACATACAAGAAACCAAGTAAAACAAA CGCATACAACGTAACGACAAATCAAGATGGC

ACAGTATCATATGGGGCTCGCCCAACACAAAACAAGCCA AGCAAAACAAATGCATATAACGTAACAACAC

ATGCAAACGGCCAAGTATCATACGGAGCTCGCCCGACAT ACAAGAAACCAAGCAAAACAAATGCATACAA

CGTAACAACACATGCAAATGGTCAAGTATCATATGGCGC TCGCCCGACACAAAAAAAGCCAAGCAAAACA

AATGCATATAACGTAACAACACATGCAAATGGTCAAGTA TCATACGGAGCTCGCCCGACATACAAGAAGC

CAAGCGAAACAAATGCATACAACGTAACAACACATGCAA ATGGTCAAGTATCATATGGAGCTCGCCCGAC

ACAAAAAAAGCCAAGCGAAACAAACGCATATAACGTAAC AACACATGCAGATGGTACTGCGACATATGGG

CCTAGAGTAACAAAATAA

Chapter 5

DETERMINATION OF MICROBIAL ACTIVITIES

INTRODUCTION

The evaluation of the microbial potential to spoil food, produce biogenic amines and produce antimicrobial activity is rarely addressed in current microbial manuals and handbooks. Therefore, this chapter initially deals with the microbial potential to spoil foods by production of sulfide compounds such as hydrogen sulfide by utilizing sulfur containing amino acids. Microbial spoilage potential is also evaluated by their ability to reduce trimethylamine oxide to trimethylamine, the typical compound which is responsible for the fishy odor in spoiled fish for instance. This chapter provides two simple and easy qualitative tests to evaluate microbial ability to produce sulfide compounds and reduce trimethylamine oxide. These tests can be used to evaluate the ability of fresh muscled foods flora during chilled storage in order to evaluate their storage stability. Biogenic amines, namely histamine, cadaverine and putrescine are responsible for scombroid food poisoning, a typical food poisoning caused by eating certain fish. These amines are produced by bacterial utilization of certain amino acids. Many studies have quantified the biogenic amines in foods. However, to the student and researcher the mechanisms of biogenic amines and responsible bacteria should be known. Therefore, this chapter provides a simple test to evaluate the ability of a bacterium to produce these biogenic amines. The test allows the student and researcher to apply the test on other biogenic amines by simply adding the amino acid which is responsible to produce the biogenic amine in question. This test can be used to avoid certain types of foods which have the potential to contain biogenic amines such as certain type of fish. Bio-preservation is a novel technology in food preservation. This technology is simply based on the ability of some bacteria to inhibit the growth of spoilage flora by producing certain compounds called bacteriocins. The characterization of bacteriocins is a complicated process that can be conducted only by a trained person. However, bacterial ability to produce these bacteriocins can be done by a simple test. This chapter trains the student and researcher through a simple test, to evaluate the microbial ability to inhibit the growth of other microorganisms by producing bacteriocins which can be assessed simply by calculating the diameter of the inhibition zone on the media. This test not only can be used in food-bio-preservation study but can also be used in controlling food poisoning if the tested microbe shows an ability to inhibit the growth of pathogens.

Determination of Microbial Activity to Produce Hydrogen Sulfide

Aims

- To understand the basics of microbial spoilage.
- To evaluate the microbial ability to spoil foods by producing hydrogen sulfide.

Introduction

Muscled foods such as meat, fish and chicken undergo different chemical, microbial and physical changes during storage which lead to food spoilage. These changes are either caused by food natural enzymes or microbial enzymes. Some of these changes can alter food sensory properties such as odor which can be detected simply by organoleptic assessment. Hydrogen sulfide produces the sulfurous odor which is considered as the typical spoilage odor in muscled foods in which food is no longer fit for consumption. Hydrogen sulfide is produced by spoilage bacteria such as *Shewanella putrefaciens* and *Aeromonas hydrophila* which utilize sulfur containing amino acids such as cysteine and methionine and produce hydrogen sulfide.

Principle

Bacteria decompose amino acids (cysteine, cystine, methionine) and reduce inorganic sulfur compounds (sulfite, thiosulfate) to hydrogen sulfide (H_2S) which can be easily detected by forming back colonies in iron agar (Gram et al. 1987) or turning the color of sulfide indole motile medium (SIM) to black due to precipitation of iron sulfate.

Materials

- Fresh colonies (24-48 hour-colonies of your purified colonies)
- 10 ml of sulfide indole motile medium (SIM)
- Needle
- Incubator at 25°C

Method

- Stab the bacterial growth into SIM medium.
- Incubate the medium at 25°C for two days.
- Observe the changes in the color of the medium.

Result

- Turing the medium to black: positive for H_2S production.
- No changes in the color of the medium: negative for H_2S production.
- Repeat the test three times to confirm the results.

Determination of Microbial Activity to Reduce Trimethylamine Oxide

Aims

- To characterize microbial spoilage.
- To evaluate the microbial ability to spoil foods.

Introduction

Trimethylamine oxide (TMAO) is a non-protein nitrogen compound found in many foods such as fish and serves as a buffering capacity agent. Trimethylamine oxide is reduced mainly by spoilage such as *Shewanella putrefaciens* to trimethylamine (TMA). Formation of TMA is associated with a fishy odor in foods such as fish. The appearance of fishy odor indicates the spoilage of muscled foods such as fish.

Principle

The microbial ability to reduce TMAO to TMA can be assessed in a medium developed by Gram et al. (1987). This medium is supplemented with TMAO which is reduced by the target bacterium, and as a result, the color of the medium changes from red to yellow.

Materials

- Fresh colonies (24-48 hour-colonies of your purified colonies)
- TMAO-medium (Gram et al. 1987):
- 2% peptone
- 0.3% Lab lemco powder
- 0.3% yeast extract
- 0.4% NaCl
- 0.4% KH_2PO_4
- 0.575% K_2HPO_4
- 0.05% $MgSO_4$
- 0.5% trimethylamine N-oxide dihydrate
- 0.2% agar
- pH: 6.8 with 2N NaOH
- Tibs 1 ml

- Incubator at 25°C
- Mineral oil

Method

Preparation of TMAO medium:

- Mix all ingredients in water.
- Adjust the pH to 6.8.
- Dissolve two tablets of resazurin per 500 ml of medium.
- Bring the medium to boiling until the tablets are dissolved.
- Dispense the medium into 8-9 ml test tubes.
- Autoclave at 115°C for 10 min.

Redution of TMAO Test

- Inoculate 1 ml aliquot of 24–48 hour–old isolates into TMAO medium.
- Immediately seal the medium with mineral oil.
- Incubate the medium 72 ± 2 h at 25°C.
- Observe the color change in TMAO medium.

Result

- Changing the color in the TMAO medium from red to yellow: positive for TMAO reduction
- No change in the TMAO medium: negative for TMAO reduction
- Repeat the test three times to confirm the results.

Determination of Microbial Activity to Hydrolyze Proteins

Aims

- To evaluate food spoilage by proteins hydrolysis.
- To enumerate proteolytic bacterial counts in foods.

Introduction

Microbial enzymes target food main components such as carbohydrates, proteins, lipids and fats. In muscled food such as meat, fish and chicken, bacteria mainly target proteins, non-protein compounds, lipids and fat. Bacterial actions on these compounds result changes in food's texture, flavor and odor. Therefore, any bacteria capable of producing these changes are called specific spoilage microorganisms. Bacteria which hydrolyze proteins such as *Pseudomonas fluorescens* and *Aeromonas hydrophila* are called proteolytic bacteria. These bacteria produce a group of enzymes called proteases which breakdown proteins. Breaking down proteins reduces the firmness of the foods which can be used an indicator of microbial spoilage.

Principle

The bacterial ability to produce protease can be assessed in any non-selective medium supplemented with a protease substrate such as gelatin (Dajanta et al. 2009). If the target bacterium produces protease, it can produce a clear hydrolysis zone on the medium after incubation.

Materials

- Pre-prepared plates of nutrient agar supplemented with 1% (w/w) gelatin. The base medium is prepared according to the instructions of the manufacturer.
- Fresh colonies (24-48 hour-colonies of your purified colonies)
- Incubator at 30°C
- Loop
- Bunsen flame

Method

- Streak bacterial growth on the nutrient agar containing gelatin plates.
- Incubate the plates at 30°C for 2 days.

Result

- Appearance of clear zones on gelatin containing medium: proteolytic activity positive.
- No clear zones: proteolytic activity negative.
- Repeat the test three times to confirm the results.

Determination of Microbial Activity to Hydrolyze Lipids

Aims

- To evaluate food spoilage by lipids hydrolysis.
- To enumerate lipolytic bacterial counts in foods.

Introduction

Lipids hydrolysis in food forms one of the major problems in the quality of oily foods such as some fish and vegetable oils during storage. This problem negatively changes the flavor of foods to off-flavor or what is called rancid flavor. Besides food natural enzymes, certain types of bacteria such as *Pseudomonas* spp. and *Flavobacterium* spp. produce extracellular lipases which hydrolyze triglycerides to free fatty acids which contribute to food off-flavor. These bacteria possess lipolytic activity and are called lipolytic bacteria which can be enumerated in a medium containing Tween.

Principle

Tweens are polyoxyethylenes of monoesters of fatty acids. Most Tweens that are used in microorganisms identification are Tween 20 (lauric acid ester), Tween 40 (palmitic acid ester), Tween 60 (stearic acid ester) and Tween 80 (oleic acid ester). Tween 20 hydrolysis is used to determine microbial lipolytic activity (Boulares et al. 2013). Released fatty acids are detectable as precipitated calcium salts.

Materials

Pre-Prepared Plates of Tween Agar 20

- Peptone: 10.0g
- Calcium chloride, hydrated: 0.10g
- Sodium chloride: 5.0g

- Tween 20: 10.0g
- Agar: 15g
- Distilled water: 1000 ml
- pH: 7.0-7.4
- Fresh colonies (24-48 hour-colonies of your purified colonies)
- Incubator at 25°C
- Loop
- Bunsen flame

Method

Preparation of Tween agar Medium and Hydrolysis of Tween 20

- Dissolve all ingredients in water.
- Mix well, check the pH and autoclave at 115°C for 10 min.
- Prepare the pre-poured plates of Tween 20 agar.
- Streak bacterial growth on the plates.
- Incubate the plates at 25°C for 1-7 days. Check the plates regularly.

Result

- Opaque zones surrounding microbial growth: lipolytic activity positive.
- No clear zones: lipolytic activity negative.
- Repeat the test three times to confirm the results.

Evaluation of Microbial Biogenic Amines Formation Potential

Aims

- To familiarize with amino acid decarboxylation tests.
- To evaluate bacterial potential for biogenic amines formation.

Introduction

Biogenic amines are chemical hazards which eliminate the safety of foods. Biogenic amines cause scombroid food poisoning generally in foods and mainly in certain type of fish. Biogenic amines such as histamine, cadaverine and putrescine are the main causes of this type of food poisoning. These amines are produced mainly by the bacterial decarboxylation of certain amino acids such as histidine to produce histamine, lysine to produce cadaverine and ornithine to produce putrescine.

Principle

The bacterial ability to decarboxylate amino acids thus producing biogenic amines can be evaluated using the HD-medium developed by Yamani and Untermann (1985). This base medium can be supplied by the amino acid in question such as histidine to evaluate bacterial ability to produce histamine, lysine to evaluate the bacterial ability to produce cadaverine and ornithine to evaluate the bacterial ability to produce putrescine. Amino acid decarboxylation can be assessed by changing the color of the medium from light green (HD-medium initial color) to violet. The decarboxylation of amino acids changes the pH of the medium to an alkaline pH shift. This results in changing the color of the medium from green to violet. Strong histamine producers such as *Morganella morganii* changes the color of the medium in from 6-8 hr.

Materials

- Fresh colonies (24-48 hour-colonies of your purified colonies)
- 10 ml HD-medium (Yamani et al. 1985):
- 2 g peptone
- 1 g Lab-Lemco powder
- 5 g NaCl
- 10 g of amino acid in question
- 10 ml of 0.1% bromcresol green solution: (0.1 g of bromcresol green in 50ml of ethanol and 50 ml of water)
- 10 ml of 0.2% chlorophenol red solution: (0.2 g chlorophenol red, 9.4 ml 0.05N NaOH and water to 100ml)
- 1000 ml distilled water
- pH 5.3
- Sterile mineral oil
- Loop
- Incubator at 25°C
- Bunsen flame

Methods

Preparation of HD-Medium

- Dissolve the solid ingredients in water, mix and add the indicators (bromcresol green first then chlorophenol red).

- Adjust the pH to 5.3 with 0.1 N NaOH.
- Dispense 10 ml into test tube.
- Autoclave at 115°C for 10 min.

Decarboxylation Activity

- Inoculate the HD-medium with bacterial growth.
- Immediately seal the test tube with 0.3 ml mineral oil.
- Incubate at 30°C for 48 hr.
- Examine the changes in the color of the medium.

Result

- Changes from green to violet: positive for amino acid decarboxylation.
- No changes: negative for amino acid decarboxylation.
- Repeat the test three times to confirm the results.

Evaluation of Microbial Antimicrobial Activity

Aims

- To understand the concept of antimicrobial activity.
- To familiarize with agar well-diffusion technique.

Introduction

In a microbial community, certain microbes dominate the flora by their high competitiveness tendencies and short generation time. Other microbes such as lactic acid bacteria dominate or even kill other microbes by producing certain organic acids such as lactic and acetic acids and some protein-nature compounds called bacteriocins. These compounds inhibit or kill other microbes such as pathogens. The modes of actions of these compounds involves altering cell wall composition, interfering with enzymes, oxidizing and stopping the synthesis of DNA, RNA, and proteins in the target microbes.

Principle

Microbial antimicrobial activity can be determined by a technique called agar diffusion well assay (Valgas et al. 2007). This assay/technique involves spreading the indicator or target microbe on agar, preparing wells of 6mm in diameter in the agar after microbial spreading and adding a cell-free of potential microbial or antimicrobial producing microbes in these wells (Nigam et al. 2012). After incubation, antimicrobial activity can be seen by clear zones around the wells on the agar which are called inhibition zones. The larger the zone diameter, the more or strong is the antimicrobial activity.

Materials

- Target/indicator microbe: for instance *Staphylococcus aureus*
- Potential/antimicrobial activity producing microbe: for instance *Lactococcus lactis*
- MRS broth. The medium is prepared according to the instructions of the manufacturer.
- Tryptone soya broth. The medium is prepared according to the instructions of the manufacturer.
- Pre-prepared plates of tryptone soya agar. The medium is prepared according to the instructions of the manufacturer.

- 9 ml MRD
- Centrifuge
- Pipettor
- 100 µl tips
- Incubators, 30°C, 37°C
- Ruler
- Sterile cork borer
- 0.20 µm sterile filter
- Syringe
- pH meter
- 1N NaOH
- McFarland barium sulfate standard 0.5

Methods

Target/Indicator Microbe Preparation

- Inoculate 10 ml tryptone soya broth with *Staphylococcus aureus* culture.
- Incubate the broth for 48 ± 2 h at 37°C.
- Dilute the bacterial growth with MRD to 10^8 (turbidity is equivalent to McFarland barium sulfate standard 0.5).

Preparation of Cell-Free Filtrate of Antimicrobial Activity Producing Microbe

- Inoculate 100 ml MRS broth with *Lactococcus lactis.*
- Incubate the broth for 72 ± 2 h at 30°C.
- Centrifuge the broth at 10,000 rpm for 5 min.
- Filter the supernatant using 0.20 µm sterile filter.

- Adjust the cell-free filtrate to pH 7.0 using 1N NaOH.

Determination of Microbial Antimicrobial Activity

- Inoculate pre-prepared plates of tryptone soya gar with 100 µl of *Staphylococcus aureus* culture.
- Spread the inoculum well over the plates.
- Leave the plates to dry.
- Using the sterile cork borer, make a well of 6 mm in diameter in the inoculated plates.
- Fill the wells with 50 µl of Lactococcus *lactis* cell-free filtrate.
- Leave the plates for about 10 min. for filtrate diffusion.
- Incubate the plates for 48 ± 2 h at 37°C.

Result

- If *Lactococcus lactis* has antimicrobial activity and produces bacteriocins, it produces a clear inhibition zone on the surface of *Staphylococcus aureus* inoculated plates. The diameter of the inhibition zone can be measured and indicates the antimicrobial activity of *Lactococcus lactis*.
- No inhibition zone indicates lacking of antimicrobial activity.

REFERENCES

Al Bulushi 2016. *Laboratory Manual.* Muscat: Sultan Qaboos University Press.

Al Bulushi, I., Poole, S., Barlow, R., Deeth, H. and Dykes, G. 2010. "Speciation of Gram-Positive Bacteria in Fresh and Ambient-stored Sub-tropical Marine Fish." *International Journal of Food Microbiology* 138: 32-38.

Al Bulushi, I. 2009. *Gram-positive Bacteria in Sub-tropical Marine Fish and Their Spoilage Potential.* PhD diss., University of Queensland.

Al Bulushi, I., Poole, S., Deeth, H. and Dykes, G. 2008. "Quantitative Assessment of Total and Gram-positive Aerobic Bacteria in Fresh and Ambient-Temperature-Stored Sub-tropical Marine Fish." *World Journal of Microbiology and Biotechnology* 24: 1867-1875.

Andrews and Hammack. 2003. *BAM: Food Sampling/Preparation of Sample Homogenate.* Accessed June 10, 2017. https://www.fda.gov/Food/FoodScienceResearch/LaboratoryMethods/ucm063335.htm.

Aznar, R. and Alarco´n, B. 2003. "PCR Detection of *Listeria monocytogenes*: A Study of Multiple Factors Affecting Sensitivity." *Journal of Applied Microbiology* 2003: 958–966.

Boulares, M., Mankai, M., Aouadhi, C., Olfa, B., Hassouna, M. 2013. "Characterization and Identification of Spoilage Psychotrophic Gram-negative Bacteria Originating from Tunisian Fresh Fish." *Annals of Microbiology* 63: 733–744.

Dajanta, K., Wongkham, S., Thirach, P., Baophoeng, P., Apichartsrangkoon, A., Santithum, P. and Chukeatirote, E. 2009. "Comparative Study of Proteolytic Activity of Protease-producing Bacteria Isolated from *Thua Nao*." *Maejo International Journal of Science and Technology* 3(02): 269-276.

Das, S., Surendran, P. and Thampuran, N. 2009. "PCR-based Detection of Enterotoxigenic Isolates of *Bacillus Cereus* from Tropical Seafood." *Indian Journal of Medical Research* 129: 316-320.

Dungu, B., Hentonb, M., Bosmana, A., Fouriea, H. and Viljoena, G. 2000. "Evaluation of Two PCR-based Procedures for Typing *Clostridium perfringens*. *Journal of the South African Veterinary Association* 71(2): 111–114.

Food Protection Services, Environmental Health Services Div. 2010. *Sampling for Microbiological Analysis* accessed June 10, 2017. http://www.bccdc.ca/resourcegallery/Documents/Guidelines%20and%20Forms/Guidelines%20and%20Manuals/EH/FPS/Food/SamplingforMicrobiologicalAnalysis2.pdf.

Gram, L., Trolle, G., and Huss, H. 1987. "Detection of Specific Spoilage Bacteria from Fish Stored at Low (0°C) and High (20°C) Temperatures." *International Journal of Food Microbiology* 4: 65-72.

Halebian, S., Harris, B., Finegold, S., and Rolfe, R. 1981. "Rapid Method that Aids in Distinguishing Gram-positive from Gram-negative Anaerobic Bacteria." *Journal of Clinical Microbiology* 13(3): 444-448.

Harrigan, W. 1998. *Laboratory Methods in Food Microbiology*, *volume 3*. California: Academic Press.

Himedia. 2015. *SIM Motility Medium, Modified* Accessed July 4, 2017. http://www.himedialabs.com/TD/M181F.pdf.

Hugh, R. and Leifson, E. 1953. "The Taxonomic Significance of Fermentative versus Oxidative Metabolism of Carbohydrates by Various Gram-negative Bacteria." *Journal of Bacteriology* 66 (1): 24 - 26.

Ibarra and Alvarado 2007. "Antimicrobial Resistance of Clinical and Environmental Strains of *Vibrio Cholerae* Isolated in Lima-Peru During Epidemics of 1991 and 1998." *The Brazilian Journal of Infectious Diseases* 11(1):100-105.

International Commission on Microbiological Specifications for Foods (ICMSF) (1986). *Microorganisms in Foods*. Toronto: University of Toronto Press.

ISO 21528-2. 2004. *Microbiology of Food and Animal Feeding Stuffs -- Horizontal Methods for the Detection and Enumeration of Enterobacteriaceae -- Part 2: Colony-Count Method.* Accessed February 15, 2017. https://www.iso.org/standard/34566.html.

ISO/DIS 6649. "Meat and Meat Products-Detection and Enumeration of *Clostridium Perfringens*." Accessed July 3, 2017. http://www.oxoid.com/UK/BLUE/prod_detail/prod_detail.asp?pr=CM0733&c=UK&lang=EN.

ISO 15214.1998. *Microbiology of Food and Animal Feeding Stuffs — Horizontal Method for the Enumeration of Mesophilic Lactic Acid Bacteria — Colony-count Technique at 30 Degrees C.* Accessed July 4, 2017. https://www.iso.org/obp/ui/#iso:std:iso:15214:ed-1:v1:en.

ISO 21528-2.2004. *Microbiology of Food and Animal Feeding Stuffs — Horizontal Methods for the Detection and Enumeration of Enterobacteriaceae, Part 2: Colony-count Method.* Accessed July 5, 2017. https://shop.austrianstandards.at/Preview.action?preview=&dokkey=186973&selectedLocale=en.

Ivanova, M., Singh, R., Dharmasena, M., Gong, C., Krastanov, A. and Jiang, X. 2014. "Rapid Identification of *Campylobacter Jejuni* from

Poultry Carcasses and Slaughtering Environment Samples by Real-time PCR." *Poultry Science* 93:1587–1597.

Kang, D. and Fung, D. 2000. "Application of Thin Agar Layer Method for Recovery of Injured *Salmonella Typhimurium*." *International Journal of Food Microbiology* 54: 127–132.

Li, R., Chiou, J., Chan, E. and Chen, S. 2016. "A Novel PCR- based Approach for Accurate Identification of *Vibrio Parahaemolyticus*." *Frontiers in Microbiology* 7: 1-5.

Lindstrom, M., Keto, R., Markkula, A., Nevas, M., Hielm, S., and Korkeala, H. 2001. "Multiplex PCR Assay for Detection and Identification of *Clostridium Botulinum* Types A, B, E, and F in Food and Fecal Material." *Applied and Environmental Microbiology* 67(12) 5694–5699.

Jay, J., Loessner, M. and Golden, D. 2005. *Modern Food Microbiology, seven edition*. New York: Springer Science and Business Media.

Malle, P., Valle, M., EB, P. and Tailliez, R. 1998. "Optimization of Culture Conditions for Enumeration of H_2S Bacteria in the Flesh of Seafish." *Journal of Rapid Methods and Automation in Microbiology* 6: 129 -141.

Martineau, F., Picard, F., Roy, P., Ouellette, M., and Bergeron, M. 1998. "Species-specific and Ubiquitous-DNA-based Assays for Rapid Identification of *staphylococcus aureus*." *Journal of Clinical Microbiology* 36(3): 618 - 623.

National Center for Biotechnology Information (NCBI). "*Staphylococcus aureus* strain PVM 715 staphylocoagulase gene, partial cds." Accessed July 6, 2017. https://www.ncbi.nlm.nih.gov/nuccore/943873224?report=fasta.

Nigam, A., Kumar, A., Madhusudan, H. and Bhola, N. 2012. "In-vitro Screening of Antibacterial Activity of Lactic Acid Bacteria against Common Enteric Pathogens." *Journal of Biomedical Sciences* 4:2, doi: 10.3823/1010.

Park, S., Kim, H., Cho, W., Kim, J., Oh, M., Kim, S., Lee, B., Ricke, S., Kim, H. 2009. "Identification of *Salmonella Enterica* Subspecies I,

Salmonella Enterica SerovarsTyphimurium, Enteritidis and Typhi Using MultiplexPCR." *FEMS Microbiology Letters* 301: 137–146.

Ray, B. and Bhunia, A. 2014. *Fundamental of Food Microbiology, fifth edition*. New York: Taylor and Francis Group LLC.

Rekha, R., Rizvi, M., Jaishree, P. 2006. "Designing and Validation of Genus-specific Primers for Human Gut Flora Study." *Electronic Journal of Biotechnology* 9 (5): 505-511.

Roberts, D. and Greenwood, M. 2003. *Practical Food Microbiology,* volume 3. Massachusetts: Blackwell Publishing.

Rosec, J. Causse, V., Cruz, B., Rauzier, J and Carnat, L. 2012. "The International Standard ISO/TS 21872–1 to Study the Occurrence of Total and Pathogenic *Vibrio Parahaemolyticus* and *Vibrio Cholerae* in Seafood: Its Improvement by Use of a Chromogenic Medium and PCR." *International Journal of Food Microbiology* 157: 189–194.

Schoeller, N. and Ingham, S. 2001. "Comparison of the Baird-Parker Agar and 3M™ Petrifilm ™ Rapid *S. Aureus* Count Plate Methods for Detection and Enumeration of *Staphylococcus Aureus*." *Food Microbiology* 18: 581-587.

Sebastia, N., El-Shenawy, M., Manes, J. and Soriano, J. 2012. "Assessment of Microbial Quality of Commercial and Homemade Tiger-nut Beverages." *Letters in Applied Microbiology* 54: 299–305.

Stewart, C. 2003. "*Staphylococcus Aureus* and Staphylococcal Enterotoxins." In *Foodborne Microorganisms of Public Health Significance,* sixth edition, edited by A. Hocking, 359-379. New South Wales: Australian Institute of Food Science and Technology Incorporated, Food Microbiology Group.

U. S. Food and Drug Administration. 2001. *Bacteriological Analytical Manual. Chapter 3: Aerobic Plate Count.* Accessed January 20, 2017. http://www.fda.gov/Food/FoodScienceResearch/Laboratory Methods/ucm063346.htm.

U. S. Food and Drug Administration. 2001. *Bacteriological Analytical Manual. Chapter 18: Yeasts, Molds and Mycotoxins.* Accessed

February 20, 2017. http://www.fda.gov/Food/FoodScience Research/LaboratoryMethods/ucm071435.htm.

U. S. Food and Drug Administration. 2001. *Bacteriological Analytical Manual Chapter 12: Staphylococcus Aureus.* Accessed February 5, 2017. http://www.fda.gov/Food/FoodScienceResearch/Laboratory Methods/ucm071429.htm.

U. S. Food and Drug Administration. 2002. *Bacteriological Analytical Manual Chapter 4: Enumeration of Escherichia Coli and the Coliform Bacteria.* Accessed February 10, 2017. http://www.fda.gov/Food/FoodScienceResearch/LaboratoryMethods/ucm064948.htm.

Valgas, C., Souza, S., Smânia, E., Smânia, Jr A. 2007. "Screening Methods to Determine Antibacterial Activity of Natural Products." *Brazilian Journal of Microbiology* 38: 369-380.

Vogel, B., Venkateswaran, K., Satomi, M. and Gram, L. 2005. " Identification of *Shewanella Baltica* as the Most Important H_2S-producing Species during Iced Storage of Danish Marine Fish." *Applied and Environmental Microbiology* 71(11): 6689-6697.

Wannet, W, Reessink, M., Brunings, H., and Maas, H. 2001. "Detection of Pathogenic *Yersinia Enterocolitica* by A rapid and Sensitive Duplex PCR Assay." *Journal of Clinical Microbiology* 39 (12): 4483–4486.

Yamani, M., and Untermann, F. 1985. "Development of Histidine Decarboxylase Medium and its Application to Detect other Amino Acid Decarboxylases." *International Journal of Food Microbiology* 2: 273-278.

APPENDIX

COMMON BACTERIAL PRIMERS

Bacteria	Primers/gene	Sequences (5′ – 3′)	Amplicon size (bp)	Reference
Staphylococcus aureus	Sa442-1 Sa442-2	AATCTTTGTCGGTACA CGATATTCTTCACG CGTAATGAGATTTCAG TAGATAATACAACA	108	Martineau et al. (1998)
Escherichia coli	ECO-1 ECO-2	GACCTCGGTTTAGTTC ACAGA CACACGCTGACGCTGA CCA	585	Wang et al. 1996, cited by Rekha et al. 2006
Salmonella Typhi	STY1599-f STY1599-r	TTACC CCACA GGAAG CACGC CTCGT TCTCT GCCGT GTGGG	203	Park et al. 2009
Salmonella Typhimurium	STM4497M2-f STM4497M2-r	AACAA CGGCT CCGGT AATGA GATTG ATGAC AAACT CTTGA TTCTG AAGAT CG	310	Park et al. 2009
Salmonella Enteritidis	IE 2L IE 3R	GGATA AGGGA TCGAT AATTG CTCAC GGACT TCCAG TTATA GTAGG TGGCC	559	Wang & Yeh (2002), cited by Park et al. 2009
Salmonella spp.	STM3098-f2 STM3098-r2	TTTGG CGGCG CAGGC GATTC GCCTC CGCCT CATCA ATCCG	423	Kim et al. (2006a), cited by Park et al. 2009

Bacteria	Primers/gene	Sequences (5′ – 3′)	Amplicon size (bp)	Reference
Listeria monocytogenes	LMA LMB	CGGAGGTTCCGCAAAAGATG CCTCCAGAGTGATCGATGTT	234	Furrer et al. (1991), cited by Aznar and Alarco´n (2003)
Campylobacter jejuni	hipOF hipOR	GTGGTCATGGAAGTGCTCCAG AAA AGCTCCTATGCTTACAACTGCT GA	131	Ivanova et al. 2014
Yersinia enterocolitica	A1 A2	TTA ATG TGT ACG CTG GGA GTG GGA GTA TTC ATA TGA AGC GTC	425	Wannet et al. 2001
Vibrio parahaemolyticus	CARB-F CARB-R	ACC(T)TTGATGGAAGATA T(C)TAACTTTCTTTGTAGTGC (A)	303	Li et al, 2016
Enterotoxigenic *Bacillus cereus*	HblA1 HblA2	GCTAATGTAGTTTCACCTGTAG CAAC AATCATGCCACTGCGTGGACAT ATAA	834	Das et al. 2009
Clostridium perfringens	Alpha-toxin	TGG CTAATGTTACTGCCGTTGATAG ATAATCCCAATCATCCCAACT ATG	324	Uzal et al. 1997, cited by Dungu et al. 2000
Clostridium perfringens	Beta-toxin	AGGAG GTTTTTTTATGAAG TCTAAA TAGCTGTTACTTTGT	196	Uzal et al. 1997, cited by Dungu et al. 2000
Clostridium perfringens	Epsilon-toxin	TACTCATACTGTG GGAACTTCGATACAAGC CTCATCTCCCATAACTGCACTA TAAT TTCC	655	Uzal et al. 1997, cited by Dungu et al. 2000
Clostridium botulinum Type A	CBMLA1 CBMLA2	AGC TAC GGA GGC AGC TAT GTT CGT ATT TGG AAA GCT GAA AAG G	782	Lindstrom et al. 2001
Clostridium botulinum Type B	CBMLB1 CBMLB2	CAG GAG AAG TGG AGC GAA AA CTT GCG CCT TTG TTT TCT TG	205	Lindstrom et al. 2001
Clostridium botulinum Type E	CBMLE1 CBMLE2	CCA AGA TTT TCA TCC GCC TA GCT ATT GAT CCA AAA CGG TGA	389	Lindstrom et al. 2001
Clostridium botulinum Type F	CBMLF1 CBMLF2	CGG CTT CAT TAG AGA ACG GA TAA CTC CCC TAG CCC CGT AT	543	Lindstrom et al. 2001

IMPORTANT FOOD MICROBIOLOGY E-SOURCES

- **Oxoid Company**: http://www.oxoid.com/UK/blue/
- **Codex Alimentarius Standards**: http://www.fao.org/fao-who-codex alimentarius/standards/en/
- **Food and Drug Administration Methods**: http://www.fda.gov/Food/FoodScienceResearch/LaboratoryMethods/ucm2006949.htm
- **ISO Standards**: http://www.iso.org/iso/products/standards/catalogue_ics_browse.htm?ICS1=07&ICS2=100&ICS3=30
- **International Commission on Microbiological Specifications for Foods (ICMSF) Methods**: http://www.icmsf.org/pdf/icmsf2.pdf
- **World Health Organization Standards**: http://www.who.int/foodsafety/areas_work/food-standard/en/
- **Official Methods of Analysis of AOAC International**: http://www. eoma.aoac.org/
- **Microbiology Laboratory Guidebook**: http://www.fsis.usda.gov/wps/portal/fsis/topics/science/laboratories-and-procedures/guidebooks-and-methods
- **Official Methods for the Microbiological Analysis of Foods:** http://www.hc-sc.gc.ca/fn-an/res-rech/analy-meth/microbio/volume1-eng.php.

INDEX

C

D

E

F

G

H

I

K

L

M

N

O

P

R

S

T

V

X

Y

Z